SI UNITS

SI UNITS

B. Chiswell and E.C.M. Grigg

University of Queensland

John Wiley & Sons Australasia Pty Ltd
SYDNEY

New York   London   Toronto

ISBN and National Library of Australia
Card Number: 0 471 15588-8.

Library of Congress Catalog Card Number: 74-139498.

Registered at the G.P.O., Sydney, for transmission
through the post as a book.

Printed at The Griffin Press, Adelaide, South Australia.

PREFACE

A great number of units of measurement of physical quantities, based on the centimetre, gramme, second (CGS), the metre, kilogramme, second (MKS) and the
imperial (English) systems, have been devised from time to time and a wide variety
of such units still exists. It is only in recent years that international organisations have
made serious attempts to reach agreement on the names and symbols and to standar-
dise the quantities of such units. The *Système International d'Unités* (SI units) provides
a standardised system with all the advantages of a common language between nations
and between the various branches of science and technology.

The introduction of SI units requires that scientists, technologists and others
become familiar not only with the new system but also with those units that have been
replaced, due to the vast quantity of scientific literature in the form of journals and
textbooks using such superseded units.

The aim of this book is to present the units, symbols and rules relating to SI
units in a concise, complete and readily comprehended manner. In addition, a number
of conversion tables have been included to allow many terms used in everyday life
and commonly used in scientific and engineering practice to be readily converted into
SI units.

TABLE OF CONTENTS

Preface   v

Section 1 Historical introduction   1

Section 2 Physical quantities   6
           Basic physical quantities and symbols   6
           Derived physical quantities and symbols   7
           Usage of the terms *specific* and *molar*   7

Section 3 Basic SI units   9
           SI as a coherent system   9
           The basic SI units—symbols and definitions.   10

Section 4 Derived SI units for physical quantities   14
           Derived SI units with special names   14
           Derived SI units without special names   15

Section 5 Prefixes for SI units   16

Section 6 Rules for use of SI units   17

Section 7 Tables of physical quantities, symbols and units   19

Section 8 Tables of conversion factors   81

Appendixes
           I Electrical and magnetic units   108

II   Expressions of concentrations of solutions   111

    Molarity   111

    Formality   111

    Normality   111

    Molality   112

    Percentage   112

III Engineering units   113

    Screw threads   113

    Technical drawings   113

    Maps and plans   113

    Modules   114

IV   Sub- and superscripts—Table 41   115

V   Recommended values of physical constants   116

SECTION I

Historical Introduction

Since man first started to require quantitative expressions for the everyday acts and things of life, the physical quantities of length, mass and time have played a part in his language and thoughts. Thus in Egypt of the sixth century B.C., the *cubit* of length, and the *mina* of weight were both widely used standard quantities. Also, although the hour lines on the sundial were a thirteenth century A.D. Arabian invention, there is little doubt that the early Egyptians had divided the period from sunrise to sunset into twelve hours of day, and the time between sunset to sunrise into twelve hours of night.

Early units of length were often derived from measurements of the human body, such as the length of the foot or the width of the palm. Units of mass were derived from inanimate objects, such as stones or grains of seed. Such units as the foot, the stone and the grain are still in use today.

There is a definite historical connection, both in nomenclature and in quantity, between systems of units currently in use and those used by the ancient Egyptian and the later Mediterranean civilisations. Thus the foot length, derived originally from the length of a man's foot, was brought by the Romans to Britain from Egypt. The inch (a thumb's breadth) was defined in the Roman system as 1/12 of a foot, and the mile was 1000 paces or double steps, a pace being equal to five Roman feet.

For the purposes of this brief historical discussion, we will concentrate on the development of the standard units in the two main world systems of units, currently used (the imperial and the metric systems), noting the strengths and weaknesses of each system. By doing so the importance of a single world-wide set of rationalised units, such as that offered in SI units, will be made clear.

The Imperial (British) System of Units

The imperial system of units has been used extensively until very recently in most parts of the British Commonwealth and in the United States of America. The last few years, however, have seen more and more Commonwealth countries adopting metric units, rather than the imperial system which is based on the standard yard length and the standard pound mass. The earliest units of measurement were often based on the physical dimensions of the human body; tradition claims that early definitions of the yard in the fifteenth century were based on the distance between a man's nose and the tip of the middle finger of his extended arm. Such a "standard" could obviously *stretch* or *shrink* as required during a commercial operation, and the standard yard is today specified as the distance between two marks engraved on two gold plugs recessed into a bar of bronze alloy of one inch cross-section and thirty-eight inches long. The bar is mounted on eight equally spaced interconnected rollers and kept at a temperature of 62°F. in the National Physical Laboratory in England. As this standard yard was made (and legalised in 1856) to replace an earlier standard which was seriously damaged when the English Houses of Parliament were burnt down in 1834, four replicas are now kept in different places as a precaution against further destruction.

The pound troy and pound avoirdupois were both introduced into England about the fourteenth century. The system of troy weight is legal only for the weighing of precious metals, while the avoirdupois pound became and is still the unit of mass used in normal commerce. The standard avoirdupois mass consists of a cylinder of platinum some 3 cm in height and slightly less than 3 cm in diameter. Like the standard yard, it is kept at the National Physical Laboratory with replicas stored at various parts of the country.

There is also another English system of expression of mass used in the compounding of medical prescriptions, namely apothecaries' weight. The only common unit of mass in all three of the above systems is the smallest, called the grain.

The imperial system is made even more confusing by the lack of uniformity in the English-speaking countries. The United States gallon (a volume measurement which is based on length units) is about 17% smaller than the British gallon, and there are marked differences in such mass measurements as the hundredweight and the ton in the two countries.

The units of time commonly used in the imperial system are similar to those of the metric system, even though they do not have a decimal base (as do other units in this latter system). Although the basic unit is now the second, it would appear that in ancient Egypt the hour was the main unit used for expressing time. The Babylonians appear to have often used a sub-division of sixty parts in their calculations and the sixty minutes of the hour (and thus, at a later date, the sixty seconds of the minute) probably owe their origin to Babylonian astrologers.

The origin of the hour undoubtedly lies in experiments of ancient Egyptian astrologers with sun clocks. These workers found that the time taken for the sun to move through equal angles of arc was constant in relation to an observer at a fixed spot on earth. Probably the smallest angle that they could readily reproduce was that of 15 degrees, which could be obtained by twice bisecting the 60 degree angle of an equilateral triangle. This 15 degree angle thus became the basic unit of the sun clock and the time taken for the sun to move through this degree of arc became the basic unit of time—the hour. At an equinox, when the sun rises at due east and sets at due west, an observer sees the sun move through 180 degrees of arc. The Egyptians thus obtained twelve hours ($12 \times 15° = 180°$) in such a day.

The Metric (Continental System)

It would seem likely that our number system is based on powers of ten, that is, it is a decimal system, because human beings have ten fingers. If this is the case, it would seem a pity that we do not have twelve fingers—as 12 is divisible by 1, 2, 3, 4 and 6, it would appear to be a more useful base for a number system than 10 which is only divisible by 1, 2 and 5.

The superiority of the present-day metric system over the imperial system lies in the fact that (apart from time units) different multiples of various physical quantities are expressed in powers of ten of the basic units. Such a system, being allied to the base of our number system, immediately removes the problem of fractions in the imperial system by making use of the decimal point. Furthermore, interconversion of units in the metric system is not a problem of long and tiresome divisions or multiplications as in the imperial system, but simply involves addition or subtraction of powers of ten.

It was Simon Stevin (1548-1620) who introduced the idea of a decimal system of units, but it was approximately some two hundred years before the metric system based on such an idea was devised primarily as a practical measure for use in industry and commerce which had shown a rapid increase in Europe after the French Revolution. In 1790 the statesman Talleyrand was appointed by the French National Assembly to remedy the lack of uniformity in weights and measures. A request was also sent to the British Parliament asking the Royal Society to co-operate with the French Academy of Sciences on devising a suitable system, but no reply was ever received. Advised by members of the French Academy of Sciences, in 1793 the French Republican Government introduced a new unit of length called the *metre* which was defined as 10^{-7} of the earth's quadrant passing through Paris. The survey of this arc was completed in 1798 and three platinum standards and several iron copies of the metre were constructed. Later when it was discovered that the quadrant had been measured inaccurately, the metre was redefined as the distance between two marks on a platinum-iridium bar when the temperature of the bar is that of melting ice.

The basic unit of mass in the metric system was the gramme (Greek—a small weight) which was defined in 1795 as the mass of a thousandth part of a cubic decimetre of water.

In the nineteenth century the metric system was adopted gradually throughout the world, and in 1897 the British Parliament passed the weights and measures (metric system) Act which allowed metric units to be used as an alternative system of weights and measures in England. However, the imperial system of units based on the standard yard and standard pound is still commonly used, even though such a system is now remarkably complex. Recent developments indicate that SI units will be the only units of measurement in the United Kingdom by 1975. Certainly if England becomes a member of the European Economic Community, metrication will be essential.

In 1875, the *Conférence Générale des Poids et Mesures* (CGPM) was constituted to control international matters concerning units of measurement. The CGPM also controls the *Comité International des Poids et Mesures* (CIPM) and the *Bureau International des Poids et Mesures* (BIPM) and the latter's laboratories at Sèvres where the standard kilogramme and the former standard metre are stored.

The principle of absolute units was proposed first in 1832 by Gauss and Weber, who put forward the Gaussian system based on the millimetre, milligramme and second. An absolute system of units can be formed from the fundamental units of mass, length and time, and in 1873 the British Association introduced the CGS system which used the centimetre, gramme and second as the fundamental units. Other metric systems in common use are the tonne, metre, second (TMS units) and the metre, kilogramme, second (MKS units).

Professor Giorgi pointed out in 1903 that the MKS system of units of mechanics could be linked with electromagnetic units by using one of the electromagnetic units as a fourth basic unit. Thus if the unit of length is the metre, the unit of mass is the kilogramme, and the unit of time is the second and the permeability of a vacuum is $4\pi \times 10^{-7}$, then practical units could be used in both electromagnetic and electrostatic systems. In 1935 the International Electrotechnical Commission (IEC) accepted Giorgi's recommendations, and in 1948 the International Conference on Weights and Measures adopted the MKS definition for the ampere. In 1950, the IEC adopted the ampere as the fourth basic unit in the MKSA (or Giorgi) system.

The CGPM in 1954 agreed on one particular form of the metric system as a suitable and preferred system, and at its 1960 meeting this system was given the name *Systéme International d'Unités* (SI), which is based on the metre, kilogramme, second, ampere, Kelvin and candela units for the basic quantities of length, mass, time, electric current, thermodynamic temperature and luminous intensity. The mole, as the basic unit of amount of substance, has also been recommended for use in SI.

The system of units known as SI is based on the MKSA system. To explain why this system was chosen rather than the CGS system we will need to refer to the concept of coherence of units. This topic is fully treated in Section 3 (pp. 9, 10). It will suffice to point out at this stage that the four-unit MKSA system is coherent for both mechanical and electrical units, while the three-unit CGS system coupled with any one electrical unit, although coherent for mechanical units, is not coherent for pre-existing derived electrical units.

An arbitrary standard is clearly not as satisfactory as an expression of a fundamental unit as is a standard derived from natural physical constants. As early as 1828 Balinet suggested that the wavelength of light be used as a natural standard unit of length. However, it was not until 1960 with the definition of SI units that the unit of length (the metre) was redefined in terms of a spectral wavelength. As will be seen in Section 3 (pp. 10-13), certain of the other basic SI units are also defined in terms of natural physical constants rather than arbitrary standards.

The current rapid movement towards world-wide standardised units not only in science (which has worked only in metric units for many years now), but in all phases of technology, engineering, agriculture, medicine and any aspect of man's work which makes use of measurement, is obviously a broad study of great interest to students of the history and philosophy of science. It will suffice at this stage to point out that the last quarter of a century has seen a rapid growth in International Business Company operations in many countries, in the growth of communications systems (which need to be interchangeable around the world) and in the grouping together of neighbouring countries under favourable trading arrangements, such as the European Economic Community. These and many other changes in the modern world have produced a growing demand, if not for a common language, at least for a common system of measurement. Thus with standardised units, a high-precision machined component of some larger machine will be interchangeable around the world; a farmer in one country will be able to compare his yields directly with other countries; a doctor will be able to use standard amounts of standardised drugs in any part of the world; and even the tourist will not feel lost with unfamiliar quantities of familiar products in a foreign country.

SECTION 2

Physical Quantities

The *Conférence Générale des Poids et Mesures* (CGPM) agreed in 1960 to adopt the *Système Internationale d'Unités*, which is abbreviated as SI in all languages. The International Organization for Standardization (ISO), which is the international authority on the standardisation of the names and symbols of physical quantities, has also endorsed the system. Most of the important international scientific organisations such as the British Standards Institute (BSI), the International Union of Pure and Applied Chemistry (IUPAC), the International Union of Pure and Applied Physics (IUPAP) and many others, are members of ISO and general agreement has been reached in recent years on the names and symbols to denote various physical quantities.

Basic Physical Quantities and Symbols

There is nothing fundamental about the choice of the number of basic quantities that are regarded as independent, or about the particular choice of the set of quantities which are to be regarded as basic. These choices are simply a matter of convenience. The particular set of seven independent physical quantities and their symbols given below in Table 1a is the one that has been chosen by international agreement. Two further dimensionless physical quantities of great use in the expression of angular quantities, also agreed upon for use in SI, are given in Table 1b.

TABLE 1(a) SYMBOLS FOR BASIC PHYSICAL QUANTITIES

Basic physical quantity	Symbol
Length	l
Mass	m
Time	t
Electric current	I
Thermodynamic temperature	T
Amount of substance	n
Luminous intensity	l_v

TABLE 1(b) SYMBOLS FOR ADDITIONAL DIMENSIONLESS BASIC PHYSICAL QUANTITIES

Dimensionless physical quantity	Symbol
Plane angle	$\alpha, \beta, \gamma, \theta$ or ϕ
Solid angle	ω or Ω

Derived Physical Quantities and Symbols

The secondary or derived physical quantities are those whose dimensions are expressed in terms of the dimensions of the basic quantities. The derivation is carried out by equations involving only multiplication, division, differentiation, and/or integration. Examples of derived physical quantities are given in Table 3.

Symbols for physical quantities are governed by the following rules.

(i) The symbol for a physical quantity should be a single letter of the Latin or Greek alphabet, and should be printed in italic (sloping) type; for example, electric current, I; electric charge, Q.

(ii) The symbol for a vector quantity should be printed in bold-faced italic type: force, $\boldsymbol{F}$; momentum, $\boldsymbol{P}$.

(iii) The symbol for a physical quantity may be modified by attaching sub- and/or superscripts. Such attachments should be printed in roman (upright) type, unless they themselves represent physical quantities. In such cases italic (sloping) type is used; e.g. charge (electrostatic), Q_E; pressure (critical), P_c.

The internationally recommended symbols for various physical quantities are given in Tables 5 to 12, while Table 41 contains a list of recommended sub- and superscripts. The number of distinctive capital and lower case letters in the Latin and Greek alphabets (eighty-eight) is not sufficient to give each physical quantity a unique characteristic symbol. However, the symbols have been chosen as far as possible to avoid clashes between the symbols used for those quantities which may possibly appear in the same context. Capital letters may be used as variants for small letters, and vice versa, provided that no ambiguity occurs, for example instead of r and r_e for internal and external radius, it is permissible to write r and R.

In all cases where a clash of symbols still occurs, an author is free to choose any appropriate symbol.

Usage of the Terms *specific* **and** *molar*

The use of the adjective *specific* before the name of an extensive physical quantity (that is, one which is dependent on the amount of substance present) is restricted to the meaning *divided by mass*.

If the extensive property is denoted by a capital letter, the corresponding specific quantity may be denoted by the corresponding lower case letter; for instance, the specific volume, v, is the volume, V, divided by the mass, m, or $v = V/m$.

The reader should note that such common terms as *specific resistance* and *specific conductance* now represent misuses of the adjective *specific*.

The adjective *molar* before the name of an extensive property is restricted to the meaning *divided by amount of substance*. The subscript m used with the symbol for the extensive property denotes the corresponding molar quantity; for example, the molar volume, V_m, is the volume, V, divided by the amount of substance, n, so $V_m = V/n$. The subscript m may be omitted if there is no risk of ambiguity.

It should be noted that *molar* does not mean *divided by the number of moles*. The term molarity should not be used to refer to the concentration of a solution when using SI units (see p. 111).

Basic SI Units

SI as a Coherent System

As pointed out in Section 1, SI is based on the MKSA rather than the CGS system of units. To explain this choice we will need to refer to the concept of coherent unit systems.

A system of units is constituted in two main sections:

(i) a small number of basic units;

(ii) any number of derived units obtained from a system of equations with proportionality constants.

For example, a volume, V (a derived quantity), may be defined in terms of length by

$$V = K \; l_1 \; l_2 \; l_3$$

where K is a constant and the l values represent lengths (a basic quantity).

A system of derived units related to a basic unit is said to be a coherent system when such derived units can be obtained from the basic unit(s) by multiplication or division without the introduction of any numerical factors (including powers of ten). It is clear that in the above example volume, V, is a derived unit coherent with the basic unit of length, l, when K = 1,

i.e. $\qquad\qquad V = l_1 \; l_2 \; l_3.$

It does not matter if length is expressed in feet and volume in cubic feet, or if length is stated in centimetres and volume in cubic centimeters, the basic and derived units in both cases will be coherent as in each case

$$V = l_1 \; l_2 \; l_3.$$

However, if length is expressed in metres and volume in cubic centimetres, then

$$V = 10^6 \; l_1 \; l_2 \; l_3$$

and the basic and derived units are no longer coherent in a system in which the unit of length is the metre.

We will illustrate this problem with a further example making reference to the quantity of force, **F**. In SI, a system based on the kilogramme mass, the metre length and the second time, the unit of force is the newton, N, which is given by:

$$N = kg \times m \times s^{-2}.$$

Other common units of force are the dyne and the poundal. In terms of basic SI units,

$$dyne = 10^{-5} kg \times m \times s^{-2}$$

and

$$poundal = 0.138\ 255\ kg \times m \times s^{-2}.$$

Thus of these three derived units of force, only the newton is coherent in SI.

The Basic SI Units—Symbols and Definitions

The number of basic units used in a system of units is a matter of convenience. For example, in mechanics it has been customary to take the three quantities, length, mass and time, as basic quantities. However, it should be realised that derived quantities that can be expressed in terms of these three basic units could also be expressed in terms of any two of these basic units. Thus the unit of length and the unit of time could be defined from the same physical phenomenon, as the two units are related by the equation:

$$c = \text{velocity of light } (m \times s^{-1}) = \frac{\text{number of oscillations per second}}{\text{number of wavelengths per metre in vacuo}}.$$

If we accept the present definition for the metre based on the krypton-86 atom (see below), then the unit of time could be defined as that period in which the radiation produced by the krypton-86 transition performs some definite number of oscillations or could be defined as the time taken by a beam of light to travel c units of length in vacuo. One could then write

$$l = c \times t$$

where $c = 3 \times 10^8$, l is in metres and t in seconds.

Although we can now define all values of time in terms of length units through the use of the constant c, units of l and of t are not coherent, and furthermore any such definition is of very limited usefulness.

Any useful system of units requires at least a small number of basic units as its fundamental building blocks. Such units will usually represent the most commonly used unit of quantity for the most commonly measured quantities. Too few basic units leads quickly to a very large complexity of derived units, while too many basic units will yield a tendency for measurements in one area to be totally unrelated to the measurements of somewhat similar type, but using different basic units.

We will illustrate the point of the need for at least a certain number of basic units and the importance of coherence, by reference to the historically contentious field of electrical and magnetic quantities and units.

The CGS system of units expresses the quantity of current in terms of the three basic units of length, mass and time, that is the centimetre, gram and second. Depending on whether the current under discussion is electrostatic or electromagnetic in origin, two different dimension systems apply in the CGS system:

(i) electrostatic current has dimension
$$(\text{length})^{\frac{3}{2}} \times (\text{mass})^{\frac{1}{2}} \times (\text{time})^{-2}$$

(ii) electromagnetic current has dimension
$$(\text{length})^{\frac{1}{2}} \times (\text{mass})^{\frac{1}{2}} \times (\text{time})^{-1}$$

To convert from derived units of electromagnetic current to derived units of electrostatic current, one multiplies the former unit's dimensions by $(\text{length}) \times (\text{time})^{-1}$, i.e. centimetre per second.

The two derived units of current are obviously not coherent and also clearly very confusing, as one is using the concept of *current* in two different ways.

SI, based on the MKSA system overcomes this problem by defining current as the fourth basic unit—the ampere—which is thus introduced on the grounds that

(i) current is a quantity very commonly measured and can thus be thought of as a basic unit,

(ii) and the confusion over electrostatic and electromagnetic current now becomes irrelevant (see Appendix I). The dimension of current in SI is the ampere.

The seven* basic units of SI are listed in Table 2a, while the two further dimensionless angular units are given in Table 2b.

TABLE 2(a) BASIC SI UNITS

Basic physical quantity	Name of basic SI unit	SI symbol
length	metre	m
mass	kilogramme	kg
time	second	s
electric current	ampere	A
thermodynamic temperature	kelvin	K
luminous intensity	candela	cd
amount of substance	mole*	mol

TABLE 2(b) ADDITIONAL DIMENSIONLESS UNITS

plane angle	radian	rad
solid angle	steradian	sr

These basic units are defined as follows.

METRE (SI symbol m)

The metre is the length equal to 1 650 763·73 wavelengths in a vacuum of the radiation associated with the transition between the levels $2p_{10}$ and $5d_5$ of the krypton-86 atom.

* The CGPM has not at present adopted the mole as a basic SI unit, but as this seventh unit has been recommended by ISO, IUPAC and IUPAP as a basic unit it is considered here as the basic unit of the amount of substance.

KILOGRAMME (*SI symbol* kg)

The kilogramme is equal to the mass of the international prototype of the kilogramme (a piece of platinum-iridium kept at the BIPM at Sèvres—see Section 1).

SECOND (*SI symbol* s)

The second is the duration of 9 192 631 770 periods of the radiation corresponding to the transition between the two hyperfine levels of the ground state of the caesium-133 atom.

AMPERE (*SI symbol* A)

The ampere is that constant current which, if maintained in two straight parallel conductors of infinite length and negligible circular cross-section which are placed one metre apart in a vacuum, would produce between these conductors a force equal to 2×10^{-7} newton* per metre of length.

KELVIN (*SI symbol* K)

The kelvin is the fraction $1/273 \cdot 16$ of the thermodynamic temperature of the triple point of ice.

CANDELA (*SI symbol* cd)

The candela is the perpendicular luminous intensity of a surface of 1/600 000 of a square metre of a black body at the temperature of freezing platinum (2 046·65 K) under a pressure of 101 325 newtons per square metre.[†]

MOLE (*SI symbol* mol)

The mole is the amount of substance which contains as many specified elementary units as there are atoms in 0·012 kilogramme of carbon-12.

Of the above seven basic SI units, only the basis of definition of the metre and the second are by any means new. Although natural phenomena have been used as the basis of definition in both these cases, any difference in the quantity specified between these and earlier definitions are only of importance in extremely accurate measurements.

The two supplementary angular units of Table 2b are defined below.

RADIAN (*SI symbol* rad)

The radian is the angle subtended at the centre of a circle by an arc on its circumference equal in length to the radius of the circle.[‡]

STERADIAN (*SI symbol* sr)

The steradian is the solid angle subtended at the centre of a sphere by a spherical

---

* The newton is a derived SI unit and will be defined in Section 4.
† This pressure of air corresponds to the non-SI unit of one atmosphere of pressure.
‡ As the circumference of a circle is equal to π times twice the radius, there will be 2π radians in a complete circle, that is one radian = $360/2\pi$ = 57·296 degrees of arc.

rea on the surface of the sphere, where the spherical area has a circular boundary and an area equal to the square of the radius of the sphere.*

All other units used in SI are derived units and will be dealt with in Section 4.

---

* As the surface area of a sphere is equal to 4π times the radius squared, the total solid angle at the centre of the sphere is equal to 4π steradians.

SECTION 4

Derived SI Units for Physical Quantities

All units, other than the seven basic (and the two supplementary) units given in Table 2, are derived from these basic units. In other words all other units can be expressed in terms of (that is, using the dimensions of) the basic units.

For example, the physical quantity of power is expressed in the derived units of the watt, W, which itself can be expressed in basic units as:

$$W = kg\, m^2\, s^{-3}$$

To be acceptable in SI, a derived unit should not only be expressible in terms of the basic units, it should also be coherent with these basic SI units. Thus in SI the quantity of magnetic flux is given in the coherent unit of the weber, Wb, where

$$Wb = kg\, m^2\, s^{-2}\, A^{-1}$$

The hitherto often used expression of magnetic flux of the maxwell, Mx, is a non-coherent unit in SI as

$$Mx = 10^{-8}\, Wb = 10^{-8}\, kg\, m^2\, s^{-2}\, A^{-1}$$

Derived SI Units with Special Names

Some derived SI units are expressions for physical quantities which are very widely used. To save the necessity of always referring to measurements of such physical quantities in terms of the basic units, such units have been given special names and symbols. A list of such SI units is given in Table 3.

TABLE 3 DERIVED SI UNITS

Physical quantity	Name of SI unit	SI symbol	Dimensions of unit derived	Dimensions of unit basic
energy	joule	J		$kg\ m^2s^{-2}$
force	newton	N	$J\ m^{-1}$	$= kg\ m\ s^{-2}$
frequency	hertz	Hz		s^{-1}
power	watt	W	$J\ s^{-1}$	$= kg\ m^2s^{-3}$
electric charge	coulomb	C		$A\ s$
electric potential difference	volt	V	$J\ C^{-1} = J\ A^{-1}s^{-1}$	$= kg\ m^2s^{-3}\ A^{-1}$
electric resistance	ohm	Ω	$V\ A^{-1}$	$= kg\ m^2s^{-3}\ A^{-2}$
electric capacitance	farad	F	$A\ s\ V^{-1} = C\ V^{-1}$	$= s^4A^2\ kg^{-1}\ m^{-2}$
magnetic flux	weber	Wb	$V\ s \quad = J\ A^{-1}$	$= kg\ m^2s^{-2}\ A^{-1}$
inductance	henry	H	$Wb\ A^{-1} = V\ A^{-1}\ s$	$= kg\ m^2\ s^{-2}\ A^{-2}$
magnetic flux density	tesla	T	$Wb\ m^{-2} = V\ S\ m^{-2}$	$= kg\ s^{-2}\ A^{-1}$
luminous flux	lumen	lm		$cd\ sr$
illumination	lux	lx	$lm\ m^{-2}$	$= cd\ sr\ m^{-2}$

Recently the CIPM has agreed to recommend to the CGPM that two further derived units with special names be included in the above list. These are the pascal and the siemens:

Physical quantity	Name of SI unit	SI symbol	Dimensions of unit derived	Dimensions of unit basic
pressure	pascal	Pa	$N\ m^{-2} = J\ m^{-3}$	$= kg\ m^{-1}\ s^{-2}$
electric conductance	siemens	S	$Ω^{-1} \quad = A\ V^{-1}$	$= s^3\ A^2\ kg^{-1}\ m^{-2}$

Derived SI Units without Special Names

Although from time to time various other commonly used units which express the amounts of often-measured physical quantities may be given special names and symbols, at present other coherent derived SI units are simply named by use of either the basic unit names only, or by a combination of basic unit names and derived unit special names. For example, the physical quantity of pressure is usually expressed in terms of force per unit square area. Thus the common SI unit name for pressure is newton per square metre, $N\ m^{-2}$. However, reference to Table 3 indicates that the newton can also be expressed as joule per metre or kilogramme metre per square second. Thus pressure in $N\ m^{-2} = J\ m^{-3} = kg\ m^{-1}\ s^{-2}$, or in other words pressure can also be expressed in joules per cubic metre or kilogramme per metre per square second. All three expressions of pressure are coherent.

A number of derived SI units without special names are listed for various physical quantities in Tables 5 to 12 in Section 7.

SECTION 5

Prefixes for SI Units

The following prefixes have been approved for use in naming decimal fractions or multiples of either basic or derived SI units.

TABLE 4 SI PREFIXES

Fraction	Prefix	Symbol
10^{-18}	atto	a
10^{-15}	femto	f
10^{-12}	pico	p
10^{-9}	nano	n
10^{-6}	micro	μ
10^{-3}	milli	m
10^{-2}	centi	c
10^{-1}	deci	d
10	deka	da
10^2	hecto	h
10^3	kilo	k
10^6	mega	M
10^9	giga	G
10^{12}	tera	T

Thus 10^{-9} m $= 1$ nm

or Gg mm^{-1} ns^{-2} $= 10^9$ g$(10^{-3}$ m$)^{-1} (10^{-9}$s$)^{-2}$
$= 10^{30}$ g m^{-1} s^{-2}

As the recommended prefixes for SI units are those that possess multiples of the power of ten of three, e.g. 10^6, 10^{-15}, etc., the use of the special prefixes for the powers of ten of -2, -1, $+1$ and $+2$ is discouraged.

SECTION 6

Rules for Use of SI Units

Certain regulations governing the methods of expression of SI units are given in different sections of this book. However, to allow ready resolution of common difficulties of expressing the units, some of these regulations are collected in this section.

Expression of Numbers

(i) The decimal point may be shown as in the following:

2.04 or 2·04 or 2,04.

(ii) A number should never commence with a decimal point, for example,

0.24 not .24.

(iii) Numbers consisting of many digits should be arranged in groups of three with a full space, not a comma, separating them:

4 032.610 42 not 4,032.610,2.

Expression of Mathematical or Algebraic Operations and Symbols

Such operations and symbols are expressed in normal ways:

addition	$x + y$
subtraction	$x - y$
multiplication	xy or $x \times y$ or $x.y$
division	x/y or $\dfrac{x}{y}$ or xy^{-1}
x is equal to y	$x = y$
x is proportional to y	$x \propto y$
factorial x	$x!$

Symbols for Units and Physical Quantities

(i) *Type format*

Abbreviations for amounts (units) of physical quantities are expressed in normal type. There are no plural endings for unit symbols, nor is a full stop used to denote an abbreviation, for instance km not kms. is the abbreviation for kilometres.

Symbols used to denote physical quantities are printed in italic (sloping) type e.g. V_p for volume at constant pressure. Symbols for vector physical quantities are shown in bold-faced italic type, e.g. moment of inertia I (see p. 7 f.)

(ii) *Algebraic rules*

Symbols for units should always be written so that they obey the normal rules of algebra. This allows such symbols to be multiplied or divided as algebraic functions without any loss in consistency of the system.

For example,
$$p = 20 \cdot 0 \text{ N m}^{-2}$$
$$= 20 \cdot 0 \text{ N/m}^2$$
Thus
$$p.\text{m.} = 20 \cdot 0 \text{ N/m}$$
or
$$p.\text{m}^2/\text{N} = 20 \cdot 0.$$

In this context it clearly pays not to use double divisions signs, e.g. kg/s/A, which are very confusing. Suitable alternative expressions are:

$$\text{kg s}^{-1} \text{ A}^{-1}$$
or
$$\text{kg/(s} \times \text{A)}$$

(iii) *Choice of suitable symbols*

Symbols for physical quantities are drawn from both the upper and lower case letters of the Greek and Latin alphabets. A large number of recommended symbols for both SI units and physical quantities are given in Section 7. Obviously, for consistency, it is better to use such symbols where possible, but in cases where there is a clash of symbols for physical quantities, an alternative symbol of a suitable type may be used. Such alternatives must be clearly specified with their first use.

(iv) *Sub- and superscripts to symbols* (see Table 41, p. 115).

These can often become unclear when they occupy more than one horizontal line of type, as in $p_{30°\text{C}}$ and are probably best shown in such cases by use of brackets:
$$p_{(30°\text{C})}.$$

(v) *Use of double prefixes*

The use of double prefixes of powers of ten before units is confusing and should be avoided: e.g. 10^{-8}A, or $10^{-2} \mu \text{A}$, not cμA.

SECTION 7

Tables of Physical Quantities, Symbols and Units

The following eight tables list physical quantities, their SI symbols and units, common multiples of their units, and notes upon the quantities.

Table 5	—	Space and time	Page 21
Table 6	—	Periodic and related phenomena	Page 24
Table 7	—	Mechanics	Page 26
Table 8	—	Heat and thermodynamics	Page 32
Table 9	—	Electricity and magnetism	Page 41
Table 10	—	Light and related electromagnetic radiation	Page 50
Table 11	—	Acoustics	Page 55
Table 12	—	Atomic, nuclear and molecular physics; physical chemistry	Page 60

NOTES ON TABLES 5-12.

Permitted Units

The permitted units in column 7 of Tables 5-12 do not follow the general rule of the SI system that all units must be made up of powers or submultiples of 10^3 from the basic units or their combinations.

Permitted units fall into three main groups:

(i) Units which are so widely used in everyday life throughout the world that they would be difficult to replace, for example the units of time such as minute, hour, day, week and year. However, their use in numerical values of physical quantities for scientific purposes will be gradually phased out.

(ii) Units which have a convenient size compared to the nearest SI unit, such as the centimetre, decimetre, are (10^2 m^2), hectare (10^4 m^2) and so on, are strictly not SI as they are not 10^3 multiples and submultiples of basic units.

Note that units such as the bar (10^5 N m^{-2}), poise $(10^{-1} \text{ kg m}^{-1} \text{ s}^{-1})$, erg (10^{-7} J) etc., all bear a decimal relationship with certain SI units, but such units will be eventually phased out.

(iii) Specialised units that affect only a very limited number of people will be progressively phased out. Examples are barn (10^{-28} m^2), curie ($3 \cdot 7 \times 10^{10}$ s^{-1}), electron volt ($1 \cdot 602 \times 10^{-19}$ J) etc.

Note the use of the litre for a volume of 1 dm^3 where high precision is not being used, and also the use of the gramme for a weight of 1 millikilogramme and the tonne for a weight of 1 kilokilogramme as no unit may have two prefixes.

In column 2 of the tables, all the symbols for quantities are in italics. When a preferred symbol and reserve symbol are given, the reserve symbol is placed in parentheses. Capital letters may be used for small letters, and vice versa, if there is no ambiguity. The symbols for units in columns 4-8 are printed in roman (upright) type, and remain unaltered in the plural.

TABLE 3   SPACE AND TIME (KINEMATICS)

List of physical quantities, their recommended symbols and SI units together with other permitted units.

Quantity	Symbol for quantity	SI unit	Symbol for SI unit	Selection of convenient recommended powers of SI unit	Other convenient powers of SI unit	Other permitted units	Remarks
plane angle	$\alpha,\beta,\gamma,\theta,$ etc.	radian	rad	mrad μrad	crad	degree(...°) minute(...') second(...")	See p. 12 for definition of radian. 1 degree (...°)$=\pi/180$ radian$=1\cdot745 \times 10^{-2}$ rad; 60 minute = 1 degree; 60 second = 1 minute. The degree, minute and second should be phased out for scientific work.
solid angle	ω,Ω	steradian	sr	msr			See p. 12 for definition of steradian.
length breadth height thickness radius diameter length of path	l b h d,s r d s	metre	m	km mm μm nm	dm cm		See p. 11 for definition of metre. The micron ($=10^{-6}$ m) is to be replaced by micrometre (μm) and millimicron ($=10^{-9}$ m) by nanometre (nm). The angstrom (Å) is to be phased out (1 Å $= 10^{-10}$ m). All units in Table 13 except the metre (and its multiples and sub-multiples) are to be discontinued. $d = 2r$
area	$A,(s)$	square metre	m²	km² mm²	hm² dm² cm²	are(a) hectare(ha)	The are ($=10^2$ m² $=$ hm²) and the hectare ($=10^4$ m²) are used solely for agrarian measurements $A = \int l \, db$

Table 5 (*continued*)

Quantity	Symbol for quantity	SI unit	Symbol for SI unit	Selection of convenient recommended powers of SI unit	Other convenient powers of SI unit	Other permitted units	Remarks
volume	$V,(v)$	cubic metre	m^3	mm^3	dm^3 cm^3	litre(l)	The CGPM redefined the litre in 1964 as $1\ l = 1\ dm^3 = 10^{-3}\ m^3$. According to the *previous* definition $1\ l = 1.000\ 028\ dm^3$. Similarly 1 hectolitre (hl) $= 10^{-1}\ m^3$, 1 millilitre (ml) $= 10^{-6}\ m^3 = 1\ cm^3$. $V = \int A\,dh$ The litre should not be used for *precision* measurements.
time	t	second	s	ks ms μs ns	hs	day(d) hour(h) minute (min)	See p. 12 for definition of second. The year is the time between two consecutive passages (in the same direction) of the sun through the earth's equatorial plane. The hour, minute and day will still be used for ordinary purposes; but their use in numerical values for physical quantities is to be avoided.
angular velocity	ω	radian per second	rad/s				Angular velocity $\omega = d\alpha/dt$

Quantity	Symbol	Unit					Remarks
velocity	u, v, w, c	metre per second	m/s	km/s mm/s	kilometre per hour	(km/h)	$u = ds/dt$. The symbols u, v, w are recommended for the components of a velocity c, when vector notation is not used. c is used for the velocity of light in vacuo.
angular acceleration	α	radian per second²	rad/s²				For rotation about a fixed axis, $\alpha = d\omega/dt$. If both ω and α are vectors, the equation applies generally.
acceleration	a	metre per second²	m/s²		gal		$a = du/dt$ for rectilinear motion. If v and a are vectors, the equation applies generally.
acceleration of free fall	g						The gal $(=10^{-2}\ \text{m/s}^2)$ and the milligal are used in geodesy.
kinematic viscosity	v	square metre per second	m²/s	mm²/s	centistoke	(cSt)	kinematic viscosity $=$ viscosity (dynamic)/density. 1 centistoke (cSt) $= 10^{-6}\ \text{m}^2/\text{s}$
diffusion coefficient	D						

TABLE 6   PERIODIC AND RELATED PHENOMENA

List of physical quantities, their recommended symbols and SI units together with other permitted units.

Quantity	Symbol for quantity	SI unit	Symbol for SI unit	Selection of convenient recommended powers of SI unit	Other convenient powers of SI unit	Other permitted units	Remarks
periodic time	T	second	s	ks ms μs ns			This is the time for one cycle. $T=1/\nu$ where ν is the frequency. Note that the minute, hour etc. should be avoided.
time constant of an exponentially varying quantity	$\tau, (T)$	second	s				If F is a function of time given by $F(t)=A+Be^{-t/\tau}$ then τ is the time constant. This may be called the relaxation time, and is the time after which the quantity would reach its limit if it maintained its initial rate of variation.
frequency	f, ν	reciprocal second or Hertz	1/s=Hz	THz GHz MHz kHz			$f=\nu=1/T$. f is mainly used in electrical technology, ν is mainly used in physics and chemistry. 1 Hz is the frequency of a periodic phenomenon of which the periodic time is 1 second.
rotational frequency	n	reciprocal second or Hertz	1/s=Hz	THz GHz MHz kHz		revolution per minute revolution per second	n is the number of revolutions per unit time. Hz=s^{-1}. See under frequency.

Quantity	Symbol	Unit			Notes
angular frequency	ω	reciprocal second or Hertz	$1/s=Hz$		$\omega=2\pi f=2\pi v$ See under frequency.
wave length	λ	metre	m	μm nm	cm	$\lambda=v/c$ The angstrom ($=10^{-10}$ m) is to be phased out. The micron ($=10^{-6}$ m) is to be replaced by micrometre (μm) and milli-micron ($=10^{-9}$ m) by nanometer (nm).
wavenumber	σ, $(\bar{v})$	reciprocal metre	m^{-1}	cm^{-1}	$\sigma=\bar{v}=1/\lambda$. $\bar{v}$ and cm^{-1} have been used extensively in spectroscopy.
circular wavenumber	k				$k=2\pi\sigma\ (=2\pi\bar{v})$
damping coefficient	δ	reciprocal second	s^{-1}		If F is a function of time given by $F(t)=A\,e^{-\delta t}\sin[2\pi(t-t_0)/T]$ then δ is the damping coefficient. Often called neper/second in telecommunication.
logarithmic decrement	Λ				$\Lambda=T\delta$ where T and δ are defined by the equation for the damping coefficient. This quantity is a *pure number*. Often called a neper in telecommunication.
attenuation coefficient	α	reciprocal metre	m^{-1}		If F is a function of distance x given by $F(x)=A\,e^{-\alpha x}\cos\beta(x-x_0)$, then α is the attenuation coefficient and β is the phase coefficient
phase coefficient	β				$\gamma=\alpha+i\beta$ Often called neper/metre in telecommunication.
propagation coefficient	γ				

TABLE 7   MECHANICS

List of physical quantities, their recommended symbols and SI units together with other permitted units.

Quantity	Symbol for quantity	SI unit	Symbol for SI unit	Selection of convenient recommended powers of SI unit	Other convenient powers of SI unit	Other permitted unit	Remarks
mass	m	kilo-gramme	kg	Mg g mg µg		tonne(t)	See p. 12 for definition of kilogramme. 1 tonne(t) = 10^3 kg. The metric carat (= 2×10^{-4} kg) was adopted in 1907 for commercial transactions in diamonds, pearls and precious stones.
amount of substance	n	mole	mol				
reduced mass	μ	kilo-gramme	kg	g			$\mu = m_1 m_2 / (m_1 + m_2)$
molar mass	M	kilo-gramme per mole	kg/mol	g/mol			$M = m/n$ i.e. mass/amount of substance.
density (mass density)	ρ	kilo-gramme per cubic metre	kg/m³	g/m³ Mg/m³	kg/dm³ g/dm³ g/cm³	kg/l g/l tonne/m³ (t/m³)	ρ = mass/volume (i.e. m/V) 1 tonne per cubic metre (t m^{-3}) = 1 kg l^{-1} = 1 g ml^{-1} = 1 g cm^{-3}. Note that the litre should not be used in precision measurements.
relative density	d						This quantity is dimensionless. It is the ratio of the density of a substance to the

...density of a reference substance under conditions that must be specified for both substances. When water is the reference substance, the imprecise name "specific gravity" has been used.

Quantity	Symbol	Name					Definition / notes
volume	$V, (v)$	cubic metre	m^3	mm^3	dm^3 cm^3	litre (l)	$1\,l = 1\ dm^3 = 10^{-3}\ m^3$. Prior to 1964 $1\,l = 1\cdot000\ 028\ dm^3$. $V = \int A\,dh$. The litre should not be used for precision measurements.
specific volume	v	cubic metre per kilogramme	m^3/kg		cm^3/g		$v =$ volume/mass (V/m)
molar volume	V_m	cubic metre per mole	m^3/mol		$m^3/kmol$	$1/mol$	$V_m = V/n$
momentum	p	kilogramme metre per second	$kg\ m/s$ $(= N\ s)$				$p =$ mass/velocity (m/u) Newton, $N = kg\ m\ s^{-2}$
moment of momentum, angular momentum	b, p_Θ, L	kilogramme metre squared per second	$kg\ m^2/s$ $(= J\ s)$	$g\ m^2/s$			The moment of momentum of a particle about a point is equal to the vector product of the radius vector from this point to the particle and the momentum of the particle. $L = r \times p$. Joule, $J = kg\ m^2\ s^{-2} = N\ m$
moment of inertia	I, J	kilogramme metre squared	$kg\ m^2$				$I_x = \int(y^2\ z^2)\,dm$

Table 7 (*continued*)

Quantity	Symbol for quantity	SI unit	Symbol for SI unit	Selection of convenient powers of SI unit recommended	Other convenient powers of SI unit	Other permitted units	Remarks
force	F	newton	N	MN kN mN μN	daN cN		$N = kg\ m\ s^{-2}$. A newton is that force, which when applied to a body having a mass of 1 kg gives it an acceleration of $1\ m\ s^{-2}$. $F = m\ d^2 r/dt^2$
weight	$G, (P, W)$	newton	N				$N = kg\ m\ s^{-2}$. For definition of newton see under force. The weight of a body is that force which, when applied to the body, gives it an acceleration equal to the local acceleration of free fall.
weight density	γ	newton per cubic metre	N/m^3				weight density = weight/volume, i.e. G/V The imprecise term "specific weight" has been used for weight density.
moment of force and bending moment	M	newton metre	$N\ m = J$	MN m kN m μN m	daN m cN m		$M = r \times F$ i.e. the moment of force about a point is equal to the vector product of the force and the radius vector from the point to any point on the line of action of the force.
torque	T	newton metre	$N\ m$ $= J$	MN m kN m μN m	daN m cN m		

quantity	symbol	unit name	unit symbol				notes
pressure and normal stress	$p,(P)$ σ	newton per square metre	N/m^2 $=Pa$	GN/m^2 MN/m^2 kN/m^2 mN/m^2 $\mu N/m^2$	daN/mm^2 cN/mm^2 N/cm^2	bar mbar(mb)	Pascal (Pa) $= N\,m^{-2} = kg\,m^{-1}\,s^{-2}$ The bar $(=10^5\,N/m^2)$, hbar $(=10^7\,N/m^2)$ mbar $(=10^2\,N/m^2)$ and μbar $(=10^{-1} N/m^2)$ are extensively used in some countries, particularly in meteorology.
and shear stress	τ						
linear strain	e,ε						These quantities are *dimensionless* Linear strain = increase in length/original length $= \Delta l/l_0$.
sheer strain	γ						
volume strain	θ						Volume strain or bulk strain = increase in volume/original volume.
Poisson's ratio	μ, v						Poisson's ratio or number = lateral contraction/elongation.
Young's modulus	E	newton per square metre	N/m^2				Young's modulus (modulus of elasticity) = normal stress/linear strain.
shear modulus	G						Shear modulus (modulus of rigidity) = shear stress/shear strain.
bulk modulus	K						Bulk modulus (modulus of compression) = pressure/volume strain.
compressibility	κ	square metre per newton	m^2/N	$= pm^2/N$			Compressibility is the reciprocal of bulk modulus $$\kappa = \frac{1}{V} \cdot \frac{dV}{dp}$$
coefficient of friction	$\mu(f)$	dimension-less					Coefficient of friction (factor of friction) = force required to move one surface over another/normal force pressing the surface together.

Table 7 (*continued*)

Quantity	Symbol for quantity	SI unit	Symbol for SI unit	Selection of convenient recommended powers of SI unit	Other convenient powers of SI unit	Other permitted units	Remarks
viscosity (dynamic)	$\eta(\mu)$	newton second per square metre	N s/m²	mN s/m²	µN s/m²	centipoise (cP)	$N\,s\,m^{-2} = kg\,m^{-1}\,s^{-1} = J\,m^{-3}\,s$
fluidity	ϕ	square metre per newton second	m²/N s	m²/mN s			$\phi = 1/\eta$
kinematic viscosity	ν	square metre per second	m²/s	mm²/s		centistoke (cSt)	Kinematic viscosity=viscosity (dynamic)/ density i.e. $\nu = \eta/\rho$ 1 centistoke (cSt) $= 10^{-6}$ m²/s
diffusion coefficient	D	square metre per second	m²/s				
surface tension	$\sigma, (\gamma)$	newton per metre	N/m	mN/m			$N\,m^{-1} = kg\,s^{-2} = J\,m^{-2}$
angle of contact	θ	radian	rad	mrad	µrad	degree(...°) minute(...') second(...")	See p. 12 for definition of radian. 1 degree $= \pi/180$ radian $= 1\cdot745 \times 10^{-2}$ rad.

quantity	symbol	unit	unit symbol	multiples	notes
work	$w, (A, W)$	joule	J	GJ, MJ, kJ, mJ	$J = N\,m = W\,s$ Work is the product of force and path.
energy	E				
potential energy	E_p, V, U, ϕ				The electrical industry uses the units W h, kW h, MW h, GW h, and TW h.
kinetic energy	E_k, T, K				kilowatt hour (kW h) $1\ kW\,h = 3\cdot6\times10^6\ J = 0\cdot160\ 2\ aJ$
					electron volt (eV) An electron acquires an energy of 1 ev when it passes through a potential of 1 volt in vacuo.
power	P	watt	W	GW, MW, kW, mW, μW	$W = kg\,m^2\,s^{-3} = J\,s^{-1}$ Power = energy/time

TABLE 8   HEAT AND THERMODYNAMICS

List of physical quantities, their recommended symbols and SI units together with other permitted units.

Quantity	Symbol for quantity	SI unit	Symbol for SI unit	Selection of convenient recommended powers of SI unit	Other convenient powers of SI unit	Other permitted units	Remarks
thermodynamic temperature, absolute temperature	$T,(\Theta)$	kelvin	K				See p. 12 for definition of kelvin.
customary temperature	t,θ					degree Celsius (°C)	The degree Fahrenheit (°F) is to be phased out. The numerical values t_C (°C), t_F (°F) and T are related by $t_C = 5/9\,(t_F-32) = T-273{\cdot}15$.
temperature interval	θ	kelvin	K			°C	The term "deg" should not be used to express temperature interval.
linear expansion coefficient	α,λ ⎱	reciprocal degree	1/K			1/°C	$\alpha = \lambda = \dfrac{1}{l}\dfrac{\mathrm{d}l}{\mathrm{d}t}$
cubic expansion coefficient	α,β,γ ⎰						$\gamma = \dfrac{1}{V}\dfrac{\mathrm{d}V}{\mathrm{d}t}$
pressure coefficient	β						β (for gases) $= \dfrac{1}{p}\dfrac{\mathrm{d}p}{\mathrm{d}t}$

quantity	symbol	unit name	symbol	other units	notes
quantity of heat	$Q,(q)$	joule	J	TJ GJ MJ kJ mJ	$J = kg\,m^2\,s^{-2}$
heat flow rate	Φ	watt	W	kW mW	$W = kg\,m^2\,s^{-3} = J\,s^{-1}$
density of heat flow rate	$q,(\phi)$	watt per square metre	W/m^2	MW/m^2 kW/m^2	Density of heat flow rate is the heat flow rate/area.
thermal conductivity	$\lambda,(k)$	watt per metre degree	$W/m\ K$	$W/m\ °C$	Thermal conductivity = density of heat flow rate/temperature gradient. $W = kg\,m^2\,s^{-3} = J\,s^{-1}$
coefficient of heat transfer	h,K,U,α	watt per square metre degree	$W/m^2\ K$	$W/m^2\ °C$	Coefficient of heat transfer = density of heat flow rate/temperature difference, i.e. q/θ. The symbols α and h are preferred for the surface coefficient of heat transfer. $W = kg\,m^2\,s^{-3} = J\,s^{-1}$
thermal diffusivity	$a,(\alpha,\chi,\kappa)$	square metre per second	m^2/s		Thermal diffusivity = thermal conductivity/(density × specific heat capacity at constant pressure).
heat capacity	C	joule per degree	J/K	kJ/K	$kJ/°C$ $J/°C$ · $C = dQ/dT$ where dQ is the small quantity of heat that increases the temperature of the system by dT. $J = kg\,m^2\,s^{-2}$

Table 8 (*continued*)

Quantity	Symbol for quantity	SI unit	Symbol for SI unit	Selection of convenient recommended powers of SI unit	Other convenient powers of SI unit	Other permitted units	Remarks
specific heat capacity	c	joule per kilogramme degree	J/kg K	kJ/kg K		kJ/kg °C J/kg °C	Specific heat capacity is the heat capacity/mass i.e. $c = C/m$. The term "specific" heat" should *not* be used. $J = kg\ m^2\ s^{-2}$
heat capacity at constant volume	C_V	joule per degree	J/K	kJ/K		kJ/°C J/°C	$C_V = (\partial U/\partial T)$ $J = kg\ m^2\ s^{-2}$
specific heat capacity at constant volume	c_v	joule per kilogramme degree	J/kg K	kJ/kg K		kJ/kg °C J/kg °C	$c_v = C_V/m$ $J = kg\ m^2\ s^{-2}$
heat capacity at constant pressure	C_p	joule per degree	J/K	kJ/K		kJ/°C J/°C	$C_p = (\partial H/\partial T)_p$ $J = kg\ m^2\ s^{-2}$
specific heat capacity at constant pressure	c_p	joule per kilogramme degree	J/kg K	kJ/kg K		kJ/kg °C J/kg °C	$c_p = C_p/m$ $J = kg\ m^2\ s^{-2}$
ratio of the specific heat capacities	γ, κ						This quantity is dimensionless. $\gamma = c_p/c_v$

Quantity	Symbol	Unit	Unit symbols		Definition
molar heat capacity	C_m	joule per degree mole	J/K mol J/K kmol	J/°C mol J/°C kmol	J = kg m² s⁻²
molar heat capacity at constant volume	$C_{V,m}$	joule per degree mole	J/K mol J/K kmol	J/°C mol J/°C kmol	$C_{V,m} = (\partial U_m/\partial T)_V$ J = kg m² s⁻²
molar heat capacity at constant pressure	$C_{p,m}$	joule per degree mole	J/K mol J/K kmol	J/°C mol J/°C kmol	$C_{p,m} = (\partial H_m/\partial T)_p$ J = kg m² s⁻²
entropy	S	joule per degree kelvin	J/K kJ/K		Provided that no irreversible change takes place, the entropy of a system is increased by dQ/T when a small quantity of heat dQ enters a system whose thermodynamic temperature is T. J = kg m² s⁻²
specific entropy	s	joule per kilogramme degree	J/kg K kJ/kg K		Specific entropy is entropy/mass (i.e. S/m) J = kg m² s⁻²
molar entropy	S_m	joule per degree mole	J/K mol kJ/K mol		Molar entropy is equal to entropy/amount of substance (i.e. S/n) J = kg m² s⁻²
internal energy	$U,(E)$	joule	J		J = kg m² s⁻². The joule is the work done when the point of application of a force of 1 newton is displaced through a distance of 1 metre in the direction of the force.
molar internal energy	$U_m,(E_m)$	joule per mole	J/mol J/kmol		See under internal energy.

Table 8 (*continued*)

Quantity	Symbol for quantity	SI unit	Symbol for SI unit	Selection of convenient recommended powers of SI unit	Other convenient powers of SI unit	Other permitted units	Remarks
specific internal energy	$u,(e)$	joule per kilogramme	J/kg	MJ/kg kJ/kg			$u = U/m$ i.e. internal energy/mass $J = kg\,m^2\,s^{-2}$
enthalpy	$H,(I)$	joule	J	kJ			See under internal energy. $H = U+pV$.
specific enthalpy	$h,(i)$	joule per kilogramme	J/kg	kJ/kg			$h = H/m$, i.e. enthalpy/mass $J = kg\,m^2\,s^{-2}$
molar enthalpy	H_m	joule per mole	J/mol				$H_m = H/n$ $J = kg\,m^2\,s^{-2}$
work function or Helmholtz free energy	A	joule	J				$A = U-TS$ $J = kg\,m^2\,s^{-2}$
specific Helmholtz free energy	a	joule per kilogramme	J/kg	kJ/kg			$a = A/m$ i.e. work function/mass $J = kg\,m^2\,s^{-2}$
molar Helmholtz free energy	A_m	joule per mole	J/mol				$A_m = A/n$ i.e. work function/amount of substance. $J = kg\,m^2\,s^{-2}$
Gibbs free energy	G	joule	J	kJ			$G = H-TS = U+pV-TS$. $J = kg\,m^2\,s^{-2}$

Quantity	Symbol	Unit name	Unit symbol	Definition / units
specific Gibbs free energy	g	joule per kilo-gramme	J/kg, kJ/kg	$g = G/m$ i.e. Gibbs free energy/mass $J = \mathrm{kg\ m^2\ s^{-2}}$
molar Gibbs free energy	G_m	joule per mole	J/mol	$G_\mathrm{m} = G/n$ i.e. Gibbs free energy/amount of substance. $J = \mathrm{kg\ m^2\ s^{-2}}$
Massieu function	J	joule per degree	J/K	$J = -A/T$ $J = \mathrm{kg\ m^2\ s^{-2}}$
specific Massieu function	j	joule per kilo-gramme degree	J/kg K	$j = J/m$ i.e. Massieu function/mass $J = \mathrm{kg\ m^2\ s^{-2}}$
Planck function	Y	joule per degree	J/K	$Y = -G/T$ $J = \mathrm{kg\ m^2\ s^{-2}}$
specific Planck function	y	joule per kilo-gramme degree	J/kg K	$y = Y/m$ i.e. Planck function/mass $J = \mathrm{kg\ m^2\ s^{-2}}$
Joule-Thomson coefficient	μ, μ_JT	cubic metre degree per joule	$\mathrm{m^3\ K\ J^{-1}}$	$\mu = (\partial T/\partial p)_H$ $\mathrm{J^{-1}\ m^3\ K = N^{-1}\ m^2\ K}$
latent heat	L	joule	J, MJ, kJ	$J = \mathrm{kg\ m^2\ s^{-2}}$ Latent heat is the quantity of heat absorbed or released in an isothermal phase change.

Table 8 (*continued*)

Quantity	Symbol for quantity	SI unit	Symbol for SI unit	Selection of convenient recommended powers of SI unit	Other convenient powers of SI unit	Other permitted units	Remarks
specific latent heat	l	joule per kilo-gramme	J/kg	MJ/kg kJ/kg			$l = L/m$ i.e. latent heat/mass $J = kg\ m^2\ s^{-2}$
chemical potential of substance B	μ_B	joule per mole	J/mol				$\mu_B = (\partial G/\partial n_B)_{T,p,n_c}$ $J = kg\ m^2\ s^{-2}$
absolute activity of substance B	λ_B						This quantity is dimensionless $\lambda_B = \exp(\mu_B/RT)$
fugacity of substance B	p_B^*, f	newton per square metre	N/m^2				$p_B^* = \lambda_B \lim_{p \to 0} (y_B p/\lambda)$
relative activity of substance B	a_B						This quantity is dimensionless. $a_B = \lambda_B/\lambda_B^\cdot$ where the superscript $\cdot$ denotes a property of a pure substance.
activity coeffi-cients of solute B:							These quantities are dimensionless.

Quantity	Symbol	Unit name	Unit symbol	Definition / Notes
(i) molality basis	γ_B			$\gamma_B = (\lambda_B/m_B)/(\lambda_B/m_B)^\infty$
(ii) concentration basis	y_B			$y_B = (\lambda_B/c_B)/(\lambda_B/c_B)^\infty$
(iii) mole fraction basis	f_B			$f_B = (\lambda_B/x_B)/(\lambda_B/x_B)^\infty$
relative activity of solute **B**	a_B			This quantity is dimensionless. $a_B = m_B\gamma_B/m^\theta$. The molality of solute B, $m_B = n_B/n_A M_A$. m^θ is the standard value of molality which is usually, but not necessarily, chosen as 1 mol kg^{-1}.
ionic strength	I	mole per kilogramme	mol/kg	$I = \frac{1}{2}\sum_i m_i z_i^2$
thermal conductivity	$\lambda,(k)$	watt per metre degree	W/m K	$W m^{-1} K^{-1} = J s^{-1} m^{-1} K$
isothermal compressibility	κ, κ_T	metre squared per newton	m²/N	$\kappa = -(\partial V/\partial p)_T \cdot 1/V$ $N = kg \ m \ s^{-2} = J \ m^{-1}$
isentropic compressibility	κ_S	metre squared per newton	m²/N	$\kappa_S = -(\partial V/\partial p)_S \cdot 1/V$. This is sometimes loosely called adiabatic compressibility. $N = kg \ m \ s^{-2} = J \ m^{-1}$

Table 8 (*continued*)

Quantity	Symbol for quantity	SI unit	Symbol for SI unit	Selection of recommended convenient powers of SI unit	Other convenient powers of SI unit	Other permitted units	Remarks
(isobaric) thermal expansivity or coefficient of thermal expansion	α	reciprocal degree	K^{-1}				$\alpha = (\partial V/\partial T)_p \cdot \; 1/V$
(isochoric) thermal pressure coefficient	β	newton per square metre degree	$N\,m^{-2}\,K^{-1}$				$\beta = (\partial p/\partial T)_V = \alpha/\kappa$ $N = kg\,m\,s^{-2} = J\,m^{-1}$

TABLE 9   ELECTRICITY AND MAGNETISM

List of physical quantities, their recommended symbols and SI units together with other permitted units.

Quantity	Symbol for quantity	SI unit	Symbol for SI unit	Selection of convenient recommended powers of SI unit	Other convenient powers of SI unit	Other permitted units	Remarks
electric current	I	ampere	A	kA mA μA nA pA			See p. 12 for definition of ampere. $I = dQ/dt$ The electric current across a surface is the time rate at which charge crosses the given surface.
electric charge, quantity of electricity	Q	coulomb	C	kC μC nC pC			$Q = \int I\, dt$ C = A s
charge density, volume density of charge	ρ	coulomb per cubic metre	C/m^3	MC/m^3 kC/m^3	C/mm^3 C/cm^3		$\rho = Q/V$ i.e. charge/volume C = A s
surface charge density	σ	coulomb per square metre	C/m^2	MC/m^2 kC/m^2	C/mm^2 C/cm^2		$\sigma = Q/A$ i.e. electric charge/surface area
electric field strength	E	volt per metre	V/m	M/V/m kV/m mV/m μV/m	V/mm V/cm		$V\,m^{-1} = kg\,m\,s^{-3}\,A^{-1} = N\,C^{-1}$ Electric field strength is the magnitude of the electric field vector. The electric field vector at a point in an electric field is the force on a stationary positive charge per unit charge.

Table 9 (*continued*)

Quantity	Symbol for quantity	SI unit	Symbol for SI unit	Selection of convenient recommended powers of SI unit	Other convenient powers of SI unit	Other permitted units	Remarks
electric potential	V, ϕ	volt	V	MV kV mV μV			$\mathrm{V = kg\ m^2\ s^{-3}\ A^{-1} = J\ A^{-1}\ s^{-1}}$
potential difference, tension	U	volt	V	MV kV mV μV			$\mathrm{V = kg\ m^2\ s^{-3}\ A^{-1} = J\ A^{-1}\ s^{-1}}$
electromotive force	E	volt	V	MV kV mV μV			$\mathrm{V = kg\ m^2\ s^{-3}\ A^{-1} = J\ A^{-1}\ s^{-1}}$
electric displacement	D	coulomb per square metre	C/m²	kC/m²	C/cm²		$\mathrm{C = A\ s}$
electric flux, flux of displacement	Ψ	coulomb	C	MC kC mC			$\mathrm{C = A\ s}$. Flux is a quantity proportional to the surface integral of the normal (perpendicular) force field intensity over a given area.

Quantity	Symbol	SI unit	Unit symbol	Sub-multiples		Notes
capacitance	C	farad	F	mF μF nF pF		$C = Q/U$ i.e. electric charge/potential difference. $F = CV^{-1} = As V^{-1} = A^2 s^4 kg^{-1} m^{-2}$
permittivity	ε	farad per metre	F/m	μF/m mF/m pF/m		$\varepsilon = D/E$ i.e. displacement/electric field strength. $F = CV^{-1} = As V^{-1} = A^2 s^4 kg^{-1} m^{-2}$
permittivity of vacuum	ε_0	farad per metre	F/m			$\varepsilon_0 = 8\cdot854\ 19 \times 10^{-12}\ F\ m^{-1}$
relative permittivity	ε_r					This quantity is dimensionless. $\varepsilon_r = \varepsilon/\varepsilon_0$. This is called the *dielectric constant*, D, when it is independent of E.
electric susceptibility	χ_e					This quantity is dimensionless. $\chi_e = \varepsilon_r - 1$
electric polarisation	P	coulomb per square metre	C/m²	MC/m² kC/m²	C/cm²	$P = D - \varepsilon_0 E$ $C = As$
electric dipole moment	$p,(p_e)$	coulomb metre	C m			C m = A s m. This is the product of one of the charges of a dipole unit by the distance separating the two dipolar charges. The dipole moment, p, is related to the torque T and the field strength E by equation $T = p \times E$
permanent dipole moment of a molecule	p,μ	coulomb metre	C m			C m = A s m See under electric dipole moment.

Table 9 (*continued*)

Quantity	Symbol for quantity	SI unit	Symbol for SI unit	Selection of convenient recommended powers of SI unit	Other convenient powers of SI unit	Other permitted units	Remarks
induced dipole moment of a molecule	p, p_i	coulomb metre	C m				C m = A s m. See under electric dipole moment.
electric polarisability of a molecule	α	coulomb metre squared per volt	C m^2/V				C = A s. The molecular polarisability is the constant of proportionality between the electrical moment of the dipole induced in a molecule and the field intensity.
electric current density	$J, j, (S)$	ampere per square metre	A/m^2	MA/m^2, kA/m^2	A/mm^2, A/cm^2		
magnetic field strength	H	ampere per metre	A/m	kA/m	A/mm, A/cm		
magnetic potential difference	U_m	ampere	A	kA, mA			
magnetomotive force	F, F_m	ampere	A	kA, mA			In this case the ampere is sometimes called ampere-turn.

Quantity	Symbol	Unit	Unit symbol	Sub-multiples	Notes
magnetic flux density, magnetic induction	$\boldsymbol{B}$	tesla	T	mT, μT, nT	$T = Wb\,m^{-2} = V\,s\,m^{-2}$
magnetic flux	Φ	weber	Wb	mWb	$Wb = V\,s$
self inductance	L	henry	H	mH, μH, nH, pH	$H = Wb\,A^{-1} = V\,s\,A^{-1} = kg\,m^2\,s^{-2}\,A^{-2}$ Self inductance is the ratio of magnetic flux linking a circuit to the flux-producing current in that circuit.
mutual inductance	M, L_{12}	henry	H	mH, μH, nH, pH	$H = Wb\,A^{-1} = V\,s\,A^{-1} = kg\,m^2\,s^{-2}\,A^{-2}$ Inductance of a circuit is the rate of increase in magnetic linkage with increase of current.
coupling coefficient	$k, (\chi, \kappa)$				These quantities are dimensionless. $k = L_{12}/(L_1 L_2)^{1/2}$
leakage coefficient	σ				$\sigma = 1 - k^2$
permeability, permeability of vacuum	μ, μ_0	henry per metre	H/m	μH/m, nH/m	$\mu = \boldsymbol{B}/\boldsymbol{H}$ i.e. magnetic flux density/magnetic field strength. $H\,m^{-1} = Wb\,A^{-1}\,m^{-1} = V\,s\,A^{-1}\,m^{-1} = kg\,m\,s^{-2}\,A^{-2}$ $\mu_0 = 1.256\,64 \times 10^{-6}\,H\,m^{-1}$
relative permeability	μ_r				This quantity is dimensionless. $\mu_r = \mu/\mu_0$
magnetic susceptibility	χ, κ				This quantity is dimensionless. $\chi_m = \mu_r - 1$

Table 9 (*continued*)

Quantity	Symbol for quantity	SI unit	Symbol for SI unit	Selection of convenient recommended powers of SI unit	Other convenient powers of SI unit	Other permitted units	Remarks
electromagnetic moment, magnetic moment	m	ampere metre squared	A m²				
magnetisation	M, H_i	ampere per metre	A/m	kA/m	A/mm		$M = \mu_0^{-1} \boldsymbol{B} - \boldsymbol{H}$
magnetic polarisation	B_i, J	tesla	T	mT			$B_i = B - \mu_0 H$ $\text{T} = \text{Wb m}^{-2} = \text{V s m}^{-2} = \text{kg s}^{-2}\,\text{A}^{-1}$
velocity of propagation of electro-magnetic waves in vacuo	c, c_0	metre per second	m/s				
resistance	R	ohm	Ω	GΩ MΩ kΩ mΩ μΩ nΩ			$\Omega = \text{V A}^{-1} = \text{kg m}^2\,\text{s}^{-3}\,\text{A}^{-2}$
resistivity	ρ	ohm metre	Ωm	GΩm MΩm	Ωcm		$\rho = E/J$ i.e. electric field strength/current density when there is no

Quantity	Symbol	Unit name	Unit symbol	Submultiples	Notes
				kΩm, mΩm, μΩm, nΩm	electromotive force in the conductor. The name specific resistance should not be used for resistivity. $\Omega = V\,A^{-1} = kg\,m^2\,s^{-3}\,A^{-2}$
conductance	G	siemens	S ($=\Omega^{-1}$)	kS, mS, μS	The siemens has been adopted by ISO and CIPM. It still awaits adoption by CGPM. $G = 1/R$. $S = \Omega^{-1} = A\,V^{-1} = kg^{-1}\,m^{-2}\,s^3\,A^2$
conductivity	$\kappa, \gamma, (\sigma)$	siemens per metre	S/m ($=\Omega^{-1}\,m^{-1}$)	MS/m, kS/m, μS/m	$\kappa = 1/\rho$ i.e. 1/resistivity; $\kappa = j/E$. Also see under conductance. The name specific conductance should not be used for conductivity.
reluctance	R, R_m	reciprocal henry	1/H		$R = U_m/\phi$ i.e. magnetic potential difference/magnetic flux. $H^{-1} = A\,V^{-1}\,s^{-1} = kg^{-1}\,m^{-2}\,s^2\,A^2$
permeance	$\Lambda, (P)$	henry	H		$\Lambda = 1/R$ i.e. 1/reluctance. $H = V\,A^{-1}\,s = kg\,m^2\,s^{-2}\,A^{-2}$
self inductance	L	henry	H		$H = V\,A^{-1}\,s = kg\,m^2\,s^{-2}\,A^{-2}$
mutual inductance	M, L_{12}	henry	H		$H = V\,A^{-1}\,s = kg\,m^2\,s^{-2}\,A^{-2}$
phase displacement	ϕ				This quantity is dimensionless.
number of turns on winding	N				These quantities are dimensionless.
number of phases	m				

Table 9 (*continued*)

Quantity	Symbol for quantity	SI unit	Symbol for SI unit	Selection of convenient recommended powers of SI unit	Other convenient powers of SI unit	Other permitted units	Remarks
number of pairs of poles	p						This quantity is dimensionless.
impedance (complex impedance)	Z	ohm	Ω	$M\Omega$ $k\Omega$ $m\Omega$			$Z = R + jX$ $\Omega = V\,A^{-1} = kg\,m^2\,s^{-3}\,A^{-2}$
modulus of impedance	$|Z|$	ohm	Ω	$M\Omega$ $k\Omega$ $m\Omega$			See under impedance.
reactance	X	ohm	Ω	$M\Omega$ $k\Omega$ $m\Omega$			See under impedance. Reactance is the imaginary part of impedance.
quality factor	Q						$Q = |X|/R$ This quantity is dimensionless.
admittance	Y	siemens	S	kS mS μS			$Y = 1/Z = 1/\text{impedance}$; $Y = G + jB$ $S = A\,V^{-1} = \Omega^{-1} = kg^{-1}\,m^{-2}\,s^3\,A^2$
modulus of admittance	$|Y|$	siemens	S	kS mS μS			See under admittance.

susceptance	B	siemens	S	kS mS µS	See under admittance.
active power	P	watt	W	TW GW MW kW µW nW	$W = V A = J s^{-1} = kg\, m^2\, s^{-3}$
apparent power	$S, (P_s)$	watt	W (= V A)		The name power factor is used for P/S. See also under active power.
reactive power	$Q, (P_q)$	watt	W	var	1 W = 1 var. The var is to be phased out. See also under active power.
Faraday constant	F	coulomb per mole	C/mol		$C = A\, s$ $F = 9{\cdot}648\,70 \times 10^4\, C\, mol^{-1}$
electromotive force of a galvanic cell	E	volt	V	mV µV	$V = kg\, m^2\, s^{-3}\, A^{-1} = J\, A^{-1}\, s^{-1}$
overpotential	η	volt	V		

For electrolyte solutions and electrochemistry see Table 12.

TABLE 10   LIGHT AND RELATED ELECTROMAGNETIC RADIATION

List of physical quantities, their recommended symbols and SI units together with other permitted units.

Quantity	Symbol for quantity	SI unit	Symbol for SI unit	Selection of convenient recommended powers of SI unit	Other convenient powers of SI unit	Other permitted units	Remarks
radiant energy	Q, Q_e	joule	J				$J = kg\,m^2\,s^{-2}$ This is the energy transmitted as electromagnetic radiation.
radiant flux, radiant power	ϕ, ϕ_e	watt	W				$W = kg\,m^2\,s^{-3} = J\,s^{-1}$ Radiant flux is the time rate of flow of radiant energy.
radiant intensity	I, I_e	watt per steradian	W/sr				$I = d\phi/d\omega$ $W = kg\,m^2\,s^{-3} = J\,s^{-1}$
radiance	L	watt per steradian square metre	W/sr m²				$L = dI/dS \cos\theta$. Radiance is the radiant flux per unit solid angle per unit of projected area of the source. $W = kg\,m^2\,s^{-3} = J\,s^{-1}$
radiant emittance	M	watt per metre squared	W/m²				$M = d\phi/dS$
irradiance	E	watt per metre squared	W/m²				$E = d\phi/dS$ i.e. the radiant flux incident per unit area of surface.

Quantity	Symbol	Unit name	Unit		Description
Planck constant	n	joule second	J s		$h = 6.625\ 6 \times 10^{...}$ J s
speed of light in a vacuum	c, c_0	metre per second	m/s		$c = 2.997\ 9 \times 10^8$ m s^{-1}
luminous intensity	I, I_v	candela	cd		See p. 12 for definition of candela. This is the ratio of the luminous flux emitted by an element of a source in an infinitesimal solid angle containing this direction, to the solid angle.
luminous flux	ϕ, ϕ_v	lumen	lm		lm = cd sr. This is the time rate of flow of light.
luminance	L, L_v	candela per metre squared	cd/m^2		Luminance is the luminous intensity per unit area.
illumination	E, E_v	lux	lx	klx	Illumination is the luminous flux falling on a surface per unit area. lx = lm m^{-2} = cd sr m^{-2}
emissivity	ε				This quantity is dimensionless. This is the ratio of the radiation emitted by a surface to the radiation emitted by a complete radiator (black body) at the same temperature and under similar conditions.
refractive index	n				This is the phase velocity of radiation in free space divided by the phase velocity of the same radiation in a specified medium. It may also be defined as the ratio of the sine of the angle of incidence to the sine of the angle of refraction. This quantity is dimensionless.

Table 10 (*continued*)

Quantity	Symbol for quantity	SI unit	Symbol for SI unit	Selection of convenient powers of SI unit	Other convenient powers of SI unit	Other permitted units	Remarks
refraction	R	cubic metre	m^3				$R = (n^2-1)V/(n^2+2)$
molar refraction	R_m	cubic metre per mole	m^3/mol				$R_m = (n^2-1)V_m/(n^2+2)$
transmittance	T						This quantity is dimensionless. $T = I/I_0$
transmittancy	T_s						This quantity is dimensionless. The common logarithm of reciprocal of transmittancy is called the absorbancy. Transmittancy is the ratio of transmittance of sample to transmittance of solvent in equivalent thickness.
absorbance (napieran), napieran extinction	B						This quantity is dimensionless. $B = \ln (1/T)$
absorbance (decadic), decadic extinction	A						This quantity is dimensionless. $A = \log_{10} (1/T)$ This was formerly called optical density.

Quantity	Symbol	Unit	Unit symbol	Definition / Notes
absorptivity (napieran), napieran absorption or extinction coefficient	b	reciprocal metre	1/m	$b = B/l$
absorptivity (decadic), decadic absorption, extinction coefficient	a	reciprocal metre	1/m	$a = A/l$
molar napierian absorptivity	k	metre squared per mole	m^2/mol	$k = B/lc$ The terms molar napierian absorption or extinction coefficient have also been used.
molar decadic absorptivity	ε	metre squared per mole	m^2/mol	$\varepsilon = A/lc$ The terms molar decadic absorption or extinction coefficient have also been used.
quantum yield	ϕ			This quantity is dimensionless. This is the number of photon-induced reactions of a specified type per photon absorbed.
angle of optical rotation	α	radian	rad, mrad, μrad, crad	See p. 12 for definition of radian. 1 degree $= \pi/180$ radian $= 1{\cdot}745 \times 10^{-2}$ rad. The degree, minute, and second should be phased out.

Table 10 (*continued*)

Quantity	Symbol for quantity	SI unit	Symbol for SI unit	Selection of convenient powers of SI unit	Other recommended convenient powers of SI unit	Other permitted units	Remarks
specific optical rotary power	α_m	radian metre squared per kilogramme	rad m²/kg	mrad m²/kg μrad m²/kg	crad m²/kg		$\alpha_m = \alpha V/ml$ Note that symbols such as $[\alpha]$ are not recommended.
molar optical rotary power	α_n	radian metre squared per mole	rad m²/mol	mrad m²/mol μrad m²/mol	crad m²/mol		$\alpha_n = \alpha V/nl = \alpha/cl$

TABLE 11 ACOUSTICS

List of physical quantities, their recommended symbols and SI units together with other permitted units.

Quantity	Symbol for quantity	SI unit	Symbol for SI unit	Selection of convenient recommended powers of SI unit	Other convenient powers of SI unit	Other permitted units	Remarks
period, periodic time	T	second	s	ms μs ns			This is the time of one cycle. Note the minute and hour should be avoided.
frequency	f, ν	reciprocal second	1/s (= Hz)				$f = \nu = 1/T$. 1Hz = 1 s^{-1}. 1 Hz is the frequency of a periodic phenomenon of which the periodic time is 1 second.
angular frequency, circular frequency	ω	reciprocal second	1/s				$\omega = 2\pi f = 2\pi \nu$
wavelength	λ	metre	m	μm nm	cm		The angstrom ($=10^{10}$ m) is to be phased out. The micron ($=10^{-6}$ m) is to be replaced by micrometer (μm) and millimicron ($=10^{-9}$ m) by nanometer (nm).
circular wave number	k	reciprocal metre	1/m				$k = \dfrac{2\pi}{\lambda}$
mass density	ρ	kilo-gramme per cubic metre	kg/m^3			g/cm^3	1 kg/m^3 = 10^{-3} g/cm^3 mass density = mass/volume

Table 11 (*continued*)

Quantity	Symbol for quantity	SI unit	Symbol for SI unit	Selection of convenient recommended powers of SI unit	Other convenient powers of SI unit	Other permitted units	Remarks
static pressure	pl	newton per square metre	$\mathrm{N/m^2}$ ($=\mathrm{Pa}$)				The static pressure is the pressure that would exist if no sound waves were present.
(instantaneous) sound pressure	p						The (instantaneous) sound pressure is the difference between the instantaneous pressure and the static pressure. $\mathrm{N = kg\,m\,s^{-2} = J\,m^{-1}} . 1\,\mathrm{Pa} = 1\,\mathrm{N\,m^{-2}}$
(instantaneous) sound particle displacement	$\xi, (x)$	metre	m	mm	cm		This is the instantaneous displacement of a particle of a medium from its mean position. The root mean square value of this quantity is often called the effective value.
(instantaneous) sound particle velocity	u, v	metre per second	m/s	mm/s	cm/s		$u = \partial\xi/\partial t$ The root mean square value of this quantity is often called the effective value.
(instantaneous) sound particle acceleration	a	metre per second squared	$\mathrm{m/s^2}$				$a = \partial u/\partial t$ The root mean square value of this quantity is often called the effective value.
(instantaneous) volume velocity	q, U	cubic metre per	$\mathrm{m^2/s}$				This is the instantaneous rate of volume flow due to sound wave across an area.

Quantity	Symbol	Unit name	Unit symbol	Description
velocity of sound	c	metre per second	m/s	The root mean square value of this quantity is often called the effective value.
sound energy density	E	joule per cubic metre	J/m^3	The sound energy density is the mean sound energy in a given volume divided by that volume. $J = kg\,m^2\,s^{-2}$
sound energy flux, sound power	$P, (N, W)$	watt	W	This is the sound energy transferred in a certain time interval divided by the duration of that interval. $W = kg\,m^2\,s^{-3} = J\,s^{-1}$ Sound power level $L_P = 10\log_{10}(P_1/P_2)$ The sound reference power must be stated. The name decibel (db) is used in telecommunication for the pure number 1.
sound intensity	I, J	watt per square metre	W/m^2	Sound intensity is the sound energy flux, (P), through an area normal to the direction of propagation divided by that area. $W = kg\,m^2\,s^{-3} = J\,s^{-1}$
acoustic impedance	$Z_a (Z)$	newton second per metre⁵	$N\,s/m^5$	The term specific acoustic impedance (Z_s) has been used when the acoustic impedance is divided by the area of surface. $N = kg\,m\,s^{-2} = J\,m^{-1}$
mechanical impedance	$Z_m, (w)$	newton second per metre	$N\,s/m$	$N = kg\,m\,s^{-2} = J\,m^{-1}$

Table 11 (*continued*)

Quantity	Symbol for quantity	SI unit	Symbol for SI unit	Selection of convenient powers of SI unit	Other recommended powers of SI unit	Other convenient units	Remarks
damping coefficient	δ	reciprocal second	1/s				In telecommunication the name neper per second has been used for reciprocal second. If F is a function of time given by $F(t) = A\,e^{-\delta t} \sin \dfrac{2\pi(t-t_0)}{T}$, then δ is the damping coefficient.
logarithmic decrement	Λ						This quantity is dimensionless. In telecommunication the term neper is often used. $\Lambda = T\delta$.
attenuation coefficient	α	reciprocal metre	1/m				If F(x) is a function of distance x then $F(x) = A\,e^{-\alpha x} \cos \beta(x-x_0)$
phase coefficient	β						In telecommunication this has often been called neper per metre
propagation coefficient	γ						$\gamma = \alpha + j\beta$
dissipation coefficient	δ						These quantities are dimensionless. δ is the ratio of the sound energy flux dissipated to the incident sound energy flux.
reflection coefficient	r, ρ						

Quantity	Symbol	Unit name	Unit symbol	Definition
transmission coefficient	τ			These quantities are dimensionless. ρ is the ratio of the sound energy flux reflected to the incident sound energy flux. τ is the ratio of the sound energy flux transmitted to the incident sound energy flux.
acoustic absorption coefficient	$\alpha, (\alpha_a)$			$\delta + \rho + \tau = 1$ $\alpha = \delta + \tau$
sound reduction index	R			$R = 10 \log_{10} 1/\tau$
equivalent absorption area of a surface or area	A	square metre	m^2	
reverberation time	T	second	s	The reverberation time is the time required for the average sound energy density in an enclosure to decrease to 10^{-6} of the initial value after the source has stopped.
loudness level	$L_N, (\Lambda)$			These quantities are dimensionless.
loudness	N			

TABLE 12   ATOMIC, NUCLEAR AND MOLECULAR PHYSICS; PHYSICAL CHEMISTRY ETC.

List of physical quantities, their recommended symbols and SI units together with other permitted units.

Quantity	Symbol for quantity	SI unit	Symbol for SI unit	Selection of convenient recommended powers of SI unit	Other convenient powers of SI unit	Other permitted units	Remarks
Avogadro constant	L, N_A	reciprocal mole	1/mol				$L = 6{\cdot}022\ 5 \times 10^{23}\ \text{mol}^{-1}$ This is the number of molecules contained in 1 mole of a substance.
gas constant	R	joule per degree per mole	J/K mol				$R = Lk$ $R = 8{\cdot}314\ 3\ \text{J K}^{-1}\ \text{mol}^{-1}$; $\text{J} = \text{kg m}^2\ \text{s}^{-2}$
Boltzmann constant	k	joule per degree	J/K				$k = 1{\cdot}380\ 5 \times 10^{-23}\ \text{J K}^{-1}$; $\text{J} = \text{kg m}^2\ \text{s}^{-2}$ $k = R/L$
Planck constant	h	joule second	J s				$h = 6{\cdot}625\ 6 \times 10^{-34}\ \text{J s}$; $\text{J} = \text{kg m}^2\ \text{s}^{-2}$ h is the energy of a quantum divided by the frequency of the radiation i.e. $h = E/\nu$
Planck constant divided by 2π	$\hbar$						$\hbar$ is the unit or quantum of orbital angular momentum. $\hbar = h/2\pi = 1{\cdot}054\ 5 \times 10^{-34}\ \text{J s}$
Rydberg constant	R_∞	reciprocal metre	1/m				The product R_c is sometimes called the Rydberg fundamental frequency.

$$R_\infty = m_e e^4/8\varepsilon_0^2 h^3 c = 1.0974 \times 10^7 \text{ m}^{-1}$$
$$R_H = R_\infty/(1+m_e/m_p) = 1.0968 \times 10^7 \text{ m}^{-1}$$

This is a quantity which has nearly the same value for all elements. If it is multiplied by a factor dependent in a regular way on the ordinal number of the line it gives the wave number of each line in a given spectral series.

name	symbol	unit		description
Faraday constant	F	coulomb per mole	C/mol	$F = Le = 9.6487 \times 10^4$ C mol^{-1} C = A s
atomic number proton number	Z Z			This quantity is dimensionless. It is the number of protons, or positive charges (expressed in terms of the electronic charge) in an atomic nucleus.
mass number nucleon number	A A			These quantities are dimensionless. The mass number represents the total number of nucleons in the nucleus, and is therefore equal to the sum of the atomic number and the neutron number. It is the whole number nearest in value to the atomic mass when that quantity is expressed in atomic mass units. The mass number should be written as a superscript before the symbol of the atom.
neutron number	N			This quantity is dimensionless. This is the number of neutrons in a nucleus. $N = A - Z$

Table 12 (*continued*)

Quantity	Symbol for quantity	SI unit	Symbol for SI unit	Selection of convenient recommended powers of SI unit	Other convenient powers of SI unit	Other permitted units	Remarks
elementary charge	e	coulomb	C				This is the unit of charge of electricity. $e = 1\cdot602\,1\times10^{-19}$ C
charge of proton	e						$C = A\,s$
rest mass of atom	m_a	kilo-gramme	kg	g			see p. 12 for definition of kilogramme.
mass of electron	m_e	kilo-gramme	kg	g			See p. 12 for definition of kilogramme. $m_e = 9\cdot109\,1\times10^{-31}$ kg
mass of proton	m_p						$m_p = 1\cdot602\,1\times10^{-19}$ kg
mass of neutron	m_n						$m_n = 1\cdot674\,8\times10^{-27}$ kg
mass of hydrogen atom	m_H						$m_H = 1\cdot673\,4\times10^{-27}$ kg
unified atomic mass constant	m_u	kilo-gramme	kg	g			$m_u = m_a\,(^{12}C)/12 = 1\cdot660\,4\times10^{-27}$ kg
molecular mass (mass of one molecule)	m	kilo-gramme	kg	g			

Quantity	Symbol	Unit name	Units	Definition
molar mass	M	kilogramme per mole	kg/mol g/mol	$M = m/n$ i.e. mass divided by amount of substance
reduced mass	μ	kilogramme	kg g	$\mu = m_1 m_2/(m_1+m_2)$
amount of substance	n	mole	mol kmol	see p. 12 for definition of mole.
molar volume	V_m	cubic metre per mole	m³/mol m³/kmol dm³/mol 1/mol	$V_m = V/n$ i.e. the volume divided by the amount of substance. The litre should not be used for precision measurements.
relative atomic mass of an element; atomic weight	A_r			This quantity is dimensionless. The atomic weight or relative atomic mass is the ratio of the average mass per atom of a specified isotopic composition of an element to 1/12 of the mass of an atom of the nuclide ^{12}C. The natural isotopic composition, unless specified otherwise, is assumed to be the specified isotopic composition.
relative molecular mass of a substance; molecular weight	M_r			This quantity is dimensionless. The molecular weight or relative molecular mass of a substance is the ratio of the average mass per molecule of a specified isotopic composition of a substance to 1/12 of the mass of an atom of the nuclide ^{12}C. The natural isotopic composition is assumed to be the specified isotopic composition unless specified otherwise.

Table 12 (*continued*)

Quantity	Symbol for quantity	SI unit	Symbol for SI unit	Selection of convenient powers of SI unit	Other convenient powers of SI unit	Other permitted units	Remarks
number of molecules	N						This quantity is dimensionless.
mole fraction of substance B	x_B, y_B						This quantity is dimensionless. The symbol y_B is used for the mole fraction of a substance B in the *gaseous* phase. $x_B = n_B / \sum_B n_B$ i.e. the number of moles of B divided by the total number of moles in the mixture.
mass fraction of substance B	ω_B						These quantities are dimensionless.
volume fraction of substance B	ϕ_B						
molality of solute B	m_B	mole per kilo-gramme	mol/kg	mol/g kmol/kg			$m_B = n_B / n_A M_A$ i.e. amount of B divided by mass of solvent. The solvent is usually denoted by symbol A and the solute by symbols B, C .... The use of the term "molal" is discouraged. See p. 12 for definition of mole.
concentration of solute B	$c_B, [B]$	mole per cubic metre	mol/m³	mol/mm³ kmol/m³	mol/dm³ mol/cm³	mol/l kmol/l	See p. 12 for definition of mole. $c_B = n_B / V$ i.e. the amount of B divided by the volume of the solution. The use of

the term "molarity" is not recommended. The use of the symbol M for mol dm^{-3} should be used only for rough values in aqueous solutions. The litre should not be used for precision measurements.

activity coefficient of solute B	γ_B y_B	This quantity is dimensionless. $\gamma_B = (\lambda_B/m_B)/(\lambda_B m_B)^\infty$ $y_B = (\lambda_B/c_B)/(\lambda_B/c_B)^\infty$ The superscript ∞ denotes the property in the limit of infinite dilution.
relative activity of solute B	a_B	This quantity is dimensionless. $a = m_B \gamma_B/m^\theta$ where m^θ is a standard value of molality usually, but not necessarily, chosen as 1 mol kg^{-1}.
relative activity of solvent A	a_A	This quantity is dimensionless. $a_A = \lambda/\lambda_A = \exp(-g M_A \sum_B m_B)$. The superscript $\cdot$ is used to denote a property of a *pure* substance.
absolute activity of substance B	μ_B	This quantity is dimensionless. $\lambda_B = \exp(\mu_B/RT)$
activity coefficient of substance B in liquid or solid mixture	f_B	This quantity is dimensionless $f_B = \lambda_B/x_B \lambda_B$. The superscript $\cdot$ denotes a property of a pure substance.

Table 12 (*continued*)

Quantity	Symbol for quantity	SI unit	Symbol for SI unit	Selection of convenient recommended powers of SI unit	Other convenient powers of SI unit	Other permitted units	Remarks
relative activity of substance B in a liquid or solid mixture	a_B						$a_B = \lambda_B/\lambda_B^{\cdot}$. The superscript $\cdot$ denotes a property of a pure substance.
osmotic coefficient of solvent A	g, ϕ						This quantity is dimensionless. $g = (M_A \sum_B m_B)^{-1} \ln(\lambda_A/\lambda_A^{\cdot})$. The superscript $\cdot$ is used to denote a property of a *pure substance*.
mass concentration of substance B	ρ_B	kilogramme per cubic metre	kg/m^3	g/m^3 kg/mm^3 g/mm^3	kg/dm^3 g/dm^3 g/cm^3	kg/l g/l	The litre should not be used for precision measurements. $\rho_B = m_B/V$ i.e. mass of B divided by the volume of the solution.
molecular concentration of substance B	c_B, n_B	reciprocal cubic metre	$1/m^3$	$1/mm^3$	$1/dm^3$ $1/cm^3$	$1/l$	$c_B = n_B = N/V$ i.e. the number of molecules or particles divided by the volume.
molecular velocity	u	metre per second	m/s		cm/s		This is a vector quantity.
average velocity	$\langle u \rangle$	metre per second	m/s		cm/s		This is a vector quantity.

Quantity	Symbol	SI unit	SI unit symbol			Other units	Notes
average speed	$\langle u \rangle$	metre per second	m/s			cm/s	
most probable speed	$\hat{u}$	metre per second	m/s			cm/s	
molecular position	$\mathbf{r}$	metre	m	mm μm nm		cm	This is a vector quantity.
molecular momentum	p	kilo-gramme metre per second	kg m/s	g m/s		g cm/s	
collision diameter of a molecule	d,σ	metre	m	μm nm			See p. 11 for definition of metre.
mean free path	l,λ	metre	m			cm	See p. 11 for definition of metre. The mean free path is the average distance that a particle or molecule travels between successive collisions with the other particles or molecules of an ensemble.
collision number	Z		1/m³ s				
grand partition function; partition function for system of prescribed T,V,μ	Ξ						This quantity is dimensionless.

Table 12 (*continued*)

Quantity	Symbol for quantity	SI unit	Symbol for SI unit	Selection of convenient powers of SI unit	Other convenient powers of SI unit	Other permitted units	Remarks
partition function for system of prescribed U,V,N	Ω						This quantity is dimensionless.
partition function for system of prescribed T,V,N	Q						This quantity is dimensionless.
partition function for a particle	q						This quantity is dimensionless.
degeneracy (multiplicity) of an energy level; statistical weight	g						This quantity is dimensionless.
symmetry number	σ,s						This quantity is dimensionless.

Quantity	Symbol	SI unit			Definition
orbital angular quantum number	L,l_i				This quantity is dimensionless.
spin angular quantum number	S,s_i				This quantity is dimensionless.
total angular quantum number	J,j_i				This quantity is dimensionless.
nuclear spin quantum number	I,J				This quantity is dimensionless.
principal quantum number	n,n_i				This quantity is dimensionless.
magnetic quantum number	M,m_i				This quantity is dimensionless.
rotational quantum number	J,K				This quantity is dimensionless.
vibrational quantum number	v				This quantity is dimensionless.
hyperfine structure quantum number	F				This quantity is dimensionless.
polarisability of a molecule	α	coulomb square metre per volt	$C\,m^2/V$	$C\,cm^2/V$	$C = As; V = kg\,m^2\,s^{-3}\,A^{-1}$ $= J\,A^{-1}\,s^{-1}$ The molecular polarisability is equal to electrical moment of the dipole induced in a molecule divided by the field intensity.

Table 12 (*continued*)

Quantity	Symbol for quantity	SI unit	Symbol for SI unit	Selection of convenient recommended powers of SI unit	Other convenient powers of SI unit	Other permitted units	Remarks
dipole moment of a molecule	p, μ	coulomb metre	C m		C cm		$C = A\,s$ The dipole moment is the product of one of the charges of a dipole unit by the distance separating the two dipolar charges. The Debye unit is to be phased out.
characteristic temperature	θ	kelvin	K				See p. 12 for definition of kelvin.
Debye temperature	θ_D	kelvin	K				See p. 12 for definition of kelvin. $\theta_D = h\nu_D/k$ i.e. the product of Planck's constant and the maximum frequency of the thermal vibrations of the lattice divided by Boltzmann constant.
Einstein temperature	θ_E	kelvin	K				See p. 12 for definition of kelvin. $\theta_E = h\nu_E/k$
rotational temperature	θ_r	kelvin	K				See p. 12 for definition of kelvin. $\theta_r = h^2/8\pi^2(I_A I_B I_C)^{1/3} k$
vibrational temperature	θ_v	kelvin	K				See p. 12 for definition of kelvin. $\theta_v = h\nu/k$
magnetic moment of a particle	μ	ampere metre squared	A m²		A cm²		

Quantity	Symbol	SI unit (name)	SI unit (symbol)	Submultiples	Other permitted units	Notes
mass excess	Δ	kilogramme	kg	g		$\Delta = m_a - Am_u$ i.e. the (rest) mass of the atom—the product of the mass number and the unified atomic mass constant.
packing fraction	f					This quantity is dimensionless. $f = \Delta/Am_u$ i.e. the mass defect divided by the product of the mass number and the unified atomic mass constant.
radioactivity; activity	A	reciprocal second (Hertz)	1/s =Hz	1/ks 1/ms 1/μs 1/ns	1/minute 1/hour	$A = -dN/dt$ i.e. the rate of disintegration of (radio) active nuclides. "Other permitted units" are to be phased out. The curie $(3.7 \times 10^{10}$ s$^{-1})$ and rutherford $(10^6$ s$^{-1})$ are to be phased out.
specific radioactivity; specific activity	a	reciprocal second reciprocal kilogramme	1/s kg	1/s g 1/s mg	1/min kg 1/min g 1/hour kg 1/hour g	"Other permitted units" are to be phased out. $a = A/m$
decay constant	λ	reciprocal second	1/s	1/ks 1/ms 1/μs 1/ns	1/min 1/hour	"Other permitted units" are to be phased out. $\lambda = -\dfrac{dN}{dt}/N = A/N$ i.e. the activity divided by the number of radioactive atoms.
half-life	$t_{1/2}$	second	s	ms ks Ms Gs	min hour year	"Other permitted units" are to be phased out. The half-life of a substance is the time required for one-half of it to undergo a disintegration or a reaction process. $t_{1/2} = \ln 2/\lambda = 0.693/\lambda$

Table 12 (*continued*)

Quantity	Symbol for quantity	SI unit	Symbol for SI unit	Selection of convenient powers of SI unit	Other convenient powers of SI unit	Other recommended permitted units	Remarks
quantum yield	ϕ						This quantity is dimensionless. This is the amount of a chemical substance divided by an amount of photons.
partial molar volume of substance B	V_B	metre cubed per mole	$m^3\ mol^{-1}$			$cm^3\ mol^{-1}$ $dm^3\ mol^{-1}$	$l\ mol^{-1}$	The litre is not to be used for precision work. $V_B = (\partial V/\partial n_B)_{T,p,n_C}$, i.e. the rate of change of V with the number of moles of B when the temperature, pressure and all other mole numbers are kept constant.
chemical potential of substance B	μ_B	joule per mole	$J\ mol^{-1}$				$J = kg\ m^2\ s^{-2}$ $\mu_B = (\partial G/\partial n_B)_{T,p,n_C}$, i.e. the rate of change of the Gibbs free energy with the number of moles of B when the temperature, pressure and all other mole numbers are kept constant.
absolute activity of substance B	λ_B						This quantity is dimensionless. $\lambda_B = \exp(\mu_B/RT)$
partial pressure of substance B in a gaseous mixture	p_B	newton per square metre	N/m^2	GN/m^2 MN/m^2 kN/m^2 mN/m^2 $\mu N/m^2$	daN/mm^2 N/cm^2		$N = kg\ m\ s^{-2} = J\ m^{-1}$ $p_B = y_B P$. The symbol y_B is used for the mole fraction of a substance B in the *gaseous* phase.

Quantity	Symbol	newton per square metre	N/m²	GN/m² MN/m² kN/m² mN/m² μN/m²	daN/mm² N/cm²	Definition / Remarks
fugacity of substance B in a gaseous mixture	p_B		N/m²		daN/mm² N/cm²	$N = kg\ m\ s^{-1} = J\ m$ $p_B^* = \lambda_B \lim_{p\to o}(y_B p/\lambda_B) =$ $y_B p \exp\left\{\int_0^p (V_B/RT - 1/p)\,dp\right\}$ The symbol y_B is used for the mole fraction of a substance B in the *gaseous* phase.
(1) equilibrium constant for a reaction in a gaseous mixture	K_{p*}			$(N/m^2)^{\Sigma_B \nu_B}$		$K_{p*} = \Pi_B(p_B^*)^{\nu_B}$. The symbol ν_B represents the stoichiometric coefficient of substance B.
(2) equilibrium constant for a reaction in a gaseous mixture	K_{p*/p^θ}					This quantity is dimensionless. $K_{p*/p^\theta} = \Pi_B(p^*/p^\theta)^{\nu_B}$. The symbol ν_B represents the stoichiometric coefficient of substance B. The superscript θ indicates a standard value of a property.
(1) equilibrium constant for a reaction in a perfect gaseous mixture	K_p			$(N/m^2)^{\Sigma_B \nu_B}$		$K_p = \Pi_B(p_B)^{\nu_B} = \Pi_B(y_B p)^{\nu_B}$. The symbol ν_B represents the stoichiometric coefficient of substance B and y_B is used for the mole fraction of substance B in the *gaseous* phase. For a perfect gaseous mixture $p_B^* = y_B p$ etc. and therefore $pV = (n_B + n_c + ...)RT$
(2) equilibrium constant for a reaction in a perfect gas mixture	K_c			$(mol/m^3)^{\Sigma_B \nu_B}$		$K_c = \Pi_B(c_B)^{\nu_B}$

Table 12 (*continued*)

Quantity	Symbol for quantity	SI unit	Symbol for SI unit	Selection of recommended powers of SI unit	Other convenient powers of SI unit	Other convenient units	Remarks
equilibrium constant for a reaction in a liquid or solid mixture	K_{xf}, K_a						This quantity is dimensionless. $K_{xf} = K_a = \Pi_B(x_B f_B)^{\nu_B} = \Pi_B(a_B)^{\nu_B}$. The symbol x_B is used for the mole fraction of substance B in the liquid or solid phase and f_B is used for the activity coefficient of substance B in the solid or liquid phase; ν_B is the stoichiometric coefficient of substance B.
equilibrium constant for a reaction in an ideal liquid or solid mixture	K_x						This quantity is dimensionless. $K_x = \Pi_B(x_B)^{\nu_B}$. The symbol x_B is used for the mole fraction of substance B in the liquid or solid phase and ν_B for the stoichiometric coefficient of substance B. For an ideal liquid mixture or ideal solid mixture $f_B = f_c = 1$.
(1) equilibrium constant for a reaction in a solution	$K_{m_B\gamma_B, a_A}$			$(\mathrm{mol/kg})^{\Sigma_B \nu_B}$		$(\mathrm{mol/g})^{\Sigma_B \nu_B}$	$K_{m_B\gamma_B, a_A} = \Pi_B(m_B\gamma_B)^{\nu_B} \cdot \exp(-\nu_A g M_A \Sigma_B m_B)$. The solvent is denoted by subscript A and the solute by subscript B. The symbols ν_B, γ_B and g are used for the

Quantity	Symbol	Definition	SI unit
(2) equilibrium constant for a reaction in a solution	K_a	stoichiometric coefficient of substance B, the activity coefficient of B, and the osmotic coefficient of solvent A respectively. This quantity is dimensionless. $K_a = \Pi_B (m_B \gamma_B / m^\theta)^{\nu_B} \cdot \exp(-\nu_A g M_A \sum_B m_B)$. The solvent is denoted by subscript A and the solute by subscript B. The symbols ν_B, γ_B and g are used for the stoichiometric coefficient of substance B, the activity coefficient of B, and the osmotic coefficient of solvent A respectively, and m^θ is a standard value of molality.	
(1) equilibrium constant for a reaction in an ideal dilute solution	K_m	$K_m = \Pi_B (m_B)^{\nu_B}$. The subscript B refers to the solute. The symbol ν_B is used for the stoichiometric coefficient of substance B. For an ideal solution $\gamma_B = \gamma_c = 1$ and $g = 1$.	$(\text{mol}/\text{kg})^{\Sigma_B \nu_B}$ $(\text{mol/g})^{\Sigma_B \nu_B}$
(2) equilibrium constant for a reaction in an ideal dilute solution	$K_{m/m\theta}$	This quantity is dimensionless. $K_{m/m\theta} = \Pi_B (m_B/m^\theta)^{\nu_B}$. The subscript B refers to the solute. The symbols ν_B and m^θ are used for the stoichiometric coefficient of substance B and the standard value of the molality respectively.	

Table 12 (*continued*)

Quantity	Symbol for quantity	SI unit	Symbol for SI unit	Selection of convenient recommended powers of SI unit	Other convenient powers of SI unit	Other permitted units	Remarks
osmotic pressure of a solution	Π	newton per square metre	N/m^2	GN/m^2 MN/m^2 kN/m^2 mN/m^2 $\mu N/m^2$	daN/mm^2 N/cm^2		The osmotic pressure of a solution is the excess pressure that must be supplied to the solution to prevent the passage into it of pure solvent through a semipermeable membrane.
stoichiometric coefficient of substance B	ν_B						This quantity is dimensionless. The stoichiometric coefficient of substance B is the number of moles of B in the balanced equation; positive for products and negative for reactants.
extent of reaction	ξ	mole	mol				$d\xi = \nu_B^{-1} dn_B$. The extent of reaction relates the changes in the number of moles of B as reaction occurs. The symbol ν_B is the number of moles of B in the balanced equation; positive for products, negative for reactants. $n_B - n_{B,0} = \nu_B \xi$. The symbol n_B refers to the amount of substance B and $n_{B,0}$ is the amount of B (e.g. at the beginning of the reaction) that sets the zero of ξ.
rate of reaction	$\dot\xi, J$	mole per second	mol/s	mol/ks mol/Ms			$\dot\xi = d\xi/dt = \nu_B^{-1} dn_B/dt$. The use of mol min^{-1} and mol hour^{-1} are to be phased out.

quantity	symbol	unit name	unit symbol	definition / notes
rate of increase of concentration of substance B	v_B, r_B	mole per cubic metre per second	$mol/(m^3\,s)$	$v_B = dc_B/dt$
activation energy of a reaction	$E, E^{\ddagger}$	joule per mole	J/mol	$J = kg\,m^2\,s^{-2}$ The activation energy is the excess energy over the ground state that must be acquired by a system in order that a particular process may occur.
rate constant of an $(n+1)$th order reaction	k, k_r		$m^{3n}/(mol^n\,s)$	
affinity of a reaction	A	joule per mole	J/mol	$J = kg\,m^2\,s^{-2}$ $A = -\sum_B v_B \mu_B$. The symbol v_B represents the stoichiometric coefficient of substance B i.e. the number of moles of B in the balanced equation; positive for products, negative for reactants.
degree of dissociation	α			This quantity is dimensionless.
ionic strength	I	mole per kilogramme	mol/kg mol/g	$I = {}^1/_2 \sum_i m_i z_i^2$
charge number of ion B	z_B			This quantity is dimensionless.
charge number of a cell reaction	z			This quantity is dimensionless.
electromotive force	E	volt	V	$V = kg\,m^2\,s^{-3}\,A^{-1} = J\,A^{-1}\,s^{-1}$

Table 12 (*continued*)

Quantity	Symbol for quantity	SI unit	Symbol for SI unit	Selection of convenient powers of SI unit	Other convenient powers of SI unit	Other permitted units	Remarks
velocity of ion B	v_B	metre per second	m/s	mm/s	cm/s		$u_B = v_B/E$ i.e. the velocity of the ion divided by the electric field strength.
electric mobility of ion B	u_B, μ_B	metre squared per second per volt	$m^2/(s\,V)$	$mm^2/(s\,V)$	$cm^2/(s\,V)$		
electrolytic conductivity	κ, σ	1/ohm metre	$1/(\Omega\,m)$	$1/(\Omega\,mm)$	$1/(\Omega\,cm)$		$\kappa = j/E$. This should *not* be called specific conductance.
molar conductivity of electrolyte	Λ	metre squared per ohm per mole	$m^2/(\Omega\,mol)$		$cm^2/(\Omega\,mol)$		The term "molar" does not follow the general rule i.e. "divided by amount of substance". It means here "divided by concentration". $\Lambda = \kappa/c$ $\Omega = kg\,m^2\,s^{-3}\,A^{-2} = V\,A^{-1}$
molar conductivity of ion B	Λ_B, λ_B	metre squared per ohm per mole	$m^2/(\Omega\,mol)$		$cm^2/(\Omega\,mol)$		The term "molar" does not follow the general rule i.e. meaning "divided by amount of substance". It means here "divided by concentration". $\Lambda_B = t_B\Lambda$ i.e. the product of the transport number of the ion and the molar conductivity of the electrolyte. $\Omega = kg\,m^2\,s^{-3}\,A^{-2} = V\,A^{-1}$

Name	Symbol	SI unit	SI unit symbol		Notes
transport number of ion B; transference number; migration number	t_B				This quantity is dimensionless. $t_B = |z_B|c_B v_B / \sum_B |z_B| c_B v_B$
activity coefficient of an electrolytic solute	$\gamma_{c,A}, \gamma_{\pm}$				This quantity is dimensionless
overpotential; overtension; overvoltage	η	volt	V	mV	$V = kg\,m^2\,s^{-3}\,A^{-1} = J\,A^{-1}\,s^{-1}$ Overpotential is the excess of observed decomposition voltage of an aqueous electrolyte over the theoretical reversible decomposition voltage.
exchange current density	j_o	ampere per metre squared	A/m^2	mA/m^2	
strength of double layer	τ	coulomb per metre	C/m	C/cm	$C = A\,s$ τ is the electric moment divided by the area.
electrokinetic potential; zeta potential	ζ	volt	V		$V = kg\,m^2\,s^{-3}\,A^{-1} = J\,A^{-1}\,s^{-1}$ The electrokinetic (zeta) potential is the difference in potential between the immovable liquid layer attached to the surface of a solid phase and the movable part of the diffuse layer in the body of the liquid.

Table 12 (*continued*)

Quantity	Symbol for quantity	SI unit	Symbol for SI unit	Selection of convenient recommended powers of SI unit	Other convenient powers of SI unit	Other permitted units	Remarks
thickness of diffusion layer	δ	metre	m	mm μm	cm		
inner electric potential	ϕ	volt	V	mV			$V = kg\,m^2\,s^{-3}\,A^{-1} = J\,A^{-1}\,s^{-1}$
outer electric potential	ψ	volt	V	mV			$V = kg\,m^2\,s^{-3}\,A^{-1} = J\,A^{-1}\,s^{-1}$
surface electric potential difference	χ	volt	V	mV			$V = kg\,m^2\,s^{-3}\,A^{-1} = J\,A^{-1}\,s^{-1}$ $\chi = \phi - \psi$
diffusion coefficient	D	metre squared per second	m^2/s	mm^2/s $\mu m^2/s$ nm^2/s	cm^2/s		This is the constant of proportionality in the Fick law. The diffusion coefficient is the rate of flow of matter (flux) per unit area for unit concentration gradient.

Other quantities of interest will be found in Tables 5-11.

SECTION 8

Conversion Factor Tables

The following twenty-six tables will facilitate ready interconversion of commonly used units and SI units for the physical quantities listed below:

Table 13	(a)	Length	Small	Page 82
	(b)		Large	Page 83
Table 14	(a)	Area	Small	Page 84
	(b)		Large	Page 85
Table 15	(a)	Volume	Small	Page 86
	(b)		Large	Page 87
	(c)		Large	Page 88
Table 16		Angle		Page 89
Table 17		Time		Page 89
Table 18		Velocity		Page 90
Table 19	(a)	Mass	Small	Page 91
	(b)		Large	Page 92
Table 20		Force		Page 93
Table 21	(a)	Pressure	Small	Page 94
	(b)		Large	Page 95
Table 22	(a)	Energy	Part 1	Page 96
	(b)		Part 2	Page 97
Table 23	(a)	Power	Part 1	Page 98
	(b)		Part 2	Page 99
Table 24		Electric charge		Page 100
Table 25		Electric current		Page 100
Table 26		Electric potential		Page 101
Table 27		Electric field strength		Page 101
Table 28		Electric resistance		Page 102
Table 29		Electric resistivity		Page 102
Table 30		Electric conductance		Page 103
Table 31		Capacitance		Page 103
Table 32		Inductance		Page 104

Table 33	Magnetic flux	Page 104
Table 34	Magnetic induction	Page 105
Table 35	Magnetic field strength	Page 105
Table 36	Magnetomotive force	Page 106
Table 37	Illumination	Page 106
Table 38	Luminance	Page 107

Notes on Tables:

1. Where there are a large number of conversion factors, the tables have been arbitrarily split into (a), (b) and sometimes (c) sub-tables.

2. The correct SI expression for the physical quantity of any table is marked by an asterisk (*), and is found as the first expression in either the horizontal or vertical columns.

TABLE 13(a) LENGTH (l)

Small

Multiply→ by↘ Obtain ↓	Metre*	Centi-metre	Micron	Nano-metre	Ang-strom	Fermi	Inch
Metre* (m)	1	10^{-2}	10^{-6}	10^{-9}	10^{-10}	10^{-15}	$2 \cdot 540 \times 10^{-2}$
Centimetre (cm)	10^{2}	1	10^{-4}	10^{-7}	10^{-8}	10^{-13}	$2 \cdot 540$
Micron (μ)	10^{6}	10^{4}	1	10^{-3}	10^{-4}	10^{-9}	$2 \cdot 540 \times 10^{4}$
Nanometre (nm)	10^{9}	10^{7}	10^{3}	1	10^{-1}	10^{-6}	$2 \cdot 540 \times 10^{7}$
Angstrom (Å)	10^{10}	10^{8}	10^{4}	10	1	10^{-5}	$2 \cdot 540 \times 10^{8}$
Fermi	10^{15}	10^{13}	10^{9}	10^{6}	10^{5}	1	$2 \cdot 540 \times 10^{13}$
Inch (in)	$3 \cdot 937 \times 10$	$3 \cdot 937 \times 10^{-1}$	$3 \cdot 937 \times 10^{-5}$	$3 \cdot 937 \times 10^{-8}$	$3 \cdot 937 \times 10^{-9}$	$3 \cdot 937 \times 10^{-14}$	1

TABLE 13(b) LENGTH (*l*)

Large

Multiply→ by↘ Obtain ↓	Metre*	Centi- metre	Kilo- metre	Inch	Foot	Yard	Rod, pole or perch	Mile
Metre* (m)	1	10^{-2}	10^3	$2 \cdot 540$ $\times 10^{-2}$	$3 \cdot 048$ $\times 10^{-1}$	$9 \cdot 144$ $\times 10^{-1}$	$5 \cdot 029$	$1 \cdot 609$ $\times 10^3$
Centimetre (cm)	10^2	1	10^5	$2 \cdot 540$	$3 \cdot 048$ $\times 10$	$9 \cdot 144$ $\times 10$	$5 \cdot 029$ $\times 10^2$	$1 \cdot 609$ $\times 10^5$
Kilometre (km)	10^{-3}	10^{-5}	1	$2 \cdot 540$ $\times 10^{-5}$	$3 \cdot 048$ $\times 10^{-4}$	$9 \cdot 144$ $\times 10^{-4}$	$5 \cdot 029$ $\times 10^{-3}$	$1 \cdot 609$
Inch (in)	$3 \cdot 937$ $\times 10$	$3 \cdot 937$ $\times 10^{-1}$	$3 \cdot 937$ $\times 10^4$	1	$1 \cdot 200$ $\times 10$	$3 \cdot 600$ $\times 10$	$1 \cdot 980$ $\times 10^2$	$6 \cdot 336$ $\times 10^4$
Foot (ft)	$3 \cdot 281$	$3 \cdot 281$ $\times 10^{-2}$	$3 \cdot 281$ $\times 10^3$	$8 \cdot 333$ $\times 10^{-2}$	1	$3 \cdot 000$	$1 \cdot 650$ $\times 10$	$5 \cdot 280$ $\times 10^3$
Yard (yd)	$1 \cdot 094$	$1 \cdot 094$ $\times 10^{-2}$	$1 \cdot 094$ $\times 10^3$	$2 \cdot 778$ $\times 10^{-2}$	$3 \cdot 330$ $\times 10^{-1}$	1	$5 \cdot 500$	$1 \cdot 760$ $\times 10^3$
Rod, pole or perch	$1 \cdot 988$	$1 \cdot 988$ $\times 10^{-3}$	$1 \cdot 988$ $\times 10^2$	$5 \cdot 050$ $\times 10^{-3}$	$6 \cdot 060$ $\times 10^{-2}$	$1 \cdot 818$ $\times 10^{-1}$	1	$3 \cdot 200$ $\times 10^2$
Mile	$6 \cdot 214$ $\times 10^{-4}$	$6 \cdot 214$ $\times 10^{-6}$	$6 \cdot 214$ $\times 10^{-1}$	$1 \cdot 578$ $\times 10^{-5}$	$1 \cdot 894$ $\times 10^{-4}$	$5 \cdot 682$ $\times 10^{-4}$	$3 \cdot 125$ $\times 10^{-3}$	1

Notes: 1 English nautical mile $= 6\,080$ feet $= 1\,853 \cdot 2$ metres.
1 International nautical mile $= 6\,076$ feet $= 1\,852$ metres.
1 chain $= 22$ yd $= 66$ ft $= 20 \cdot 12$ metres.
1 furlong $= 10$ chains $= 220$ ft $= 201 \cdot 2$ metres.
1 light year $= 3 \cdot 105 \times 10^{16}$ ft $= 5 \cdot 879 \times 10^{12}$ miles $= 9 \cdot 461 \times 10^{15}$ metres.
1 parsec $= 1 \cdot 916 \times 10^{13}$ miles $= 3 \cdot 084 \times 10^{16}$ metres.

TABLE 14(a) AREA (l^2)

Small

Multiply→ by↘ Obtain ↓	Square metre*	Square centimetre	Square millimetre	Square inch	Square foot	Square yard
Square metre* (m^2)	1	1×10^4	1×10^6	$6 \cdot 452 \times 10^{-4}$	$9 \cdot 290 \times 10^{-2}$	$8 \cdot 361 \times 10^{-1}$
Square centimetre (cm)^2	1×10^{-4}	1	1×10^2	$6 \cdot 452$	$9 \cdot 290 \times 10^2$	$8 \cdot 361 \times 10^3$
Square millimetre (mm)^2	1×10^{-6}	1×10^{-2}	1	$6 \cdot 452 \times 10^2$	$9 \cdot 290 \times 10^4$	$8 \cdot 361 \times 10^5$
Square inch (in)^2	$1 \cdot 550 \times 10^3$	$1 \cdot 550 \times 10^{-1}$	$1 \cdot 550 \times 10^{-3}$	1	$1 \cdot 440 \times 10^2$	$1 \cdot 296 \times 10^3$
Square foot (ft)^2	$1 \cdot 0764 \times 10$	$1 \cdot 0764 \times 10^{-3}$	$1 \cdot 0764 \times 10^{-5}$	$6 \cdot 944 \times 10^{-3}$	1	$9 \cdot 000$
Square yard (yd)^2	$1 \cdot 1960$	$1 \cdot 1960 \times 10^{-4}$	$1 \cdot 1960 \times 10^{-6}$	$7 \cdot 692 \times 10^{-4}$	$1 \cdot 111 \times 10^{-1}$	1

ABLE 14(b) AREA (l^2)

arge

Multiply→ by↘ Obtain ↓	Square metre*	Are	Hectare	Square perch	Square chain	Acre	Square mile
quare metre*	1	1×10^2	1×10^4	$1 \cdot 619 \times 10^4$	$4 \cdot 0469 \times 10^2$	$4 \cdot 0469 \times 10^3$	$2 \cdot 590 \times 10^6$
re	1×10^{-2}	1	1×10^2	$1 \cdot 619 \times 10^2$	$4 \cdot 0469$	$4 \cdot 0469 \times 10$	$2 \cdot 590 \times 10^4$
Iectare	1×10^{-4}	1×10^{-2}	1	$1 \cdot 619$	$4 \cdot 0469 \times 10^{-2}$	$4 \cdot 0469 \times 10^{-1}$	$2 \cdot 590 \times 10^2$
quare perch	$6 \cdot 173 \times 10^{-5}$	$6 \cdot 173 \times 10^{-3}$	$6 \cdot 173 \times 10^{-1}$	1	$1 \cdot 600 \times 10$	$1 \cdot 600 \times 10^2$	$1 \cdot 024 \times 10^5$
quare chain (ch^2)	$2 \cdot 471 \times 10^{-3}$	$2 \cdot 471 \times 10^{-1}$	$2 \cdot 471 \times 10$	$6 \cdot 250 \times 10^{-2}$	1	$1 \cdot 000 \times 10$	$6 \cdot 400 \times 10^3$
cre	$2 \cdot 471 \times 10^{-4}$	$2 \cdot 471 \times 10^{-2}$	$2 \cdot 471$	$6 \cdot 250 \times 10^{-3}$	$1 \cdot 000 \times 10^{-1}$	1	$6 \cdot 400 \times 10^2$
quare mile	$3 \cdot 861 \times 10^{-7}$	$3 \cdot 861 \times 10^{-5}$	$3 \cdot 861 \times 10^{-3}$	$9 \cdot 804 \times 10^{-6}$	$1 \cdot 563 \times 10^{-4}$	$4 \cdot 096 \times 10^{-3}$	1

TABLE 15(a) VOLUME (l^3)

Small

Multiply → by ↘ Obtain ↓	Cubic metre*	Cubic decimetre (litre)	Cubic centimetre	Cubic millimetre	Cubic inch	Cubic foot	Cubic yard
Cubic metre* (m^3)	1	$1 \cdot 000 \times 10^{-3}$	$1 \cdot 000 \times 10^{-6}$	$1 \cdot 000 \times 10^{-9}$	$1 \cdot 639 \times 10^{-5}$	$2 \cdot 832 \times 10^{-2}$	$7 \cdot 646 \times 10^{-1}$
Cubic decimetre (dm^3) (litre)[a]	$1 \cdot 000 \times 10^{3}$	1	$1 \cdot 000 \times 10^{-3}$	$1 \cdot 000 \times 10^{-6}$	$1 \cdot 639 \times 10^{-2}$	$2 \cdot 832 \times 10$	$7 \cdot 646 \times 10^{2}$
Cubic centimetre (cm^3) (millilitre)[a]	$1 \cdot 000 \times 10^{6}$	$1 \cdot 000 \times 10^{3}$	1	$1 \cdot 000 \times 10^{-3}$	$1 \cdot 639 \times 10$	$2 \cdot 832 \times 10^{4}$	$7 \cdot 646 \times 10^{5}$
Cubic milli- metre (mm^3)	$1 \cdot 000 \times 10^{9}$	$1 \cdot 000 \times 10^{6}$	$1 \cdot 000 \times 10^{3}$	1	$1 \cdot 639 \times 10^{4}$	$2 \cdot 832 \times 10^{7}$	$7 \cdot 646 \times 10^{8}$
Cubic inch (in^3)	$6 \cdot 102 \times 10^{4}$	$6 \cdot 102 \times 10$	$6 \cdot 102 \times 10^{-2}$	$6 \cdot 102 \times 10^{-5}$	1	$1 \cdot 728 \times 10^{3}$	$4 \cdot 666 \times 10^{4}$
Cubic foot (ft^3)	$3 \cdot 531 \times 10$	$3 \cdot 531 \times 10^{-2}$	$3 \cdot 531 \times 10^{-5}$	$3 \cdot 531 \times 10^{-8}$	$5 \cdot 787 \times 10^{-4}$	1	$2 \cdot 700 \times 10$
Cubic yard (yd^3)	$1 \cdot 308$	$1 \cdot 308 \times 10^{-3}$	$1 \cdot 308 \times 10^{-6}$	$1 \cdot 308 \times 10^{-9}$	$2 \cdot 143 \times 10^{-5}$	$3 \cdot 704 \times 10^{-2}$	1

Note: [a] The litre is equivalent to $1 \cdot 000\,027$ dm^3. Unless extremely accurate values are being measured the litre and dm^3 can be treated as identical as can also cm^3 and millilitre.

TABLE 15(b) VOLUME (l^3)

Large[a]

Multiply→ by ↘ Obtain ↓	Cubic metre*	Litre	U.S. pint	U.K. pint	U.S. gallon	U.K. gallon	U.K. heck
Cubic metre (m³)*	1	$1{\cdot}000 \times 10^{-3}$	$4{\cdot}732 \times 10^{-4}$	$5{\cdot}683 \times 10^{-4}$	$3{\cdot}785 \times 10^{-3}$	$4{\cdot}546 \times 10^{-3}$	$9{\cdot}092 \times 10^{-3}$
Litre[b] (l)	$1{\cdot}000 \times 10^{3}$	1	$4{\cdot}732 \times 10^{-1}$	$5{\cdot}683 \times 10^{-1}$	$3{\cdot}785$	$4{\cdot}546$	$9{\cdot}092$
U.S. pint[c] (pt)	$2{\cdot}113 \times 10^{3}$	$2{\cdot}113$	1	$1{\cdot}201$	$8{\cdot}000$	$9{\cdot}608$	$1{\cdot}922 \times 10$
U.K. pint[c] (pt)	$1{\cdot}760 \times 10^{3}$	$1{\cdot}760$	$8{\cdot}327 \times 10^{-1}$	1	$6{\cdot}662$	$8{\cdot}000$	$1{\cdot}600 \times 10$
U.S. gallon[e] (gal)	$2{\cdot}642 \times 10^{2}$	$2{\cdot}642 \times 10^{-1}$	$1{\cdot}250 \times 10^{-1}$	$1{\cdot}501 \times 10^{-1}$	1	$1{\cdot}201$	$2{\cdot}402$
U.K. gallon[e] (gal)	$2{\cdot}200 \times 10^{2}$	$2{\cdot}200 \times 10^{-1}$	$1{\cdot}041 \times 10^{-1}$	$1{\cdot}250 \times 10^{-1}$	$8{\cdot}327 \times 10^{-1}$	1	$2{\cdot}000$
U.K. heck[d]	$1{\cdot}100 \times 10^{2}$	$1{\cdot}100 \times 10^{-1}$	$5{\cdot}205 \times 10^{-2}$	$6{\cdot}125 \times 10^{-2}$	$4{\cdot}164 \times 10^{-1}$	$5{\cdot}000 \times 10^{-1}$	1

Notes: [a] These volumes are all essentially liquid measures.
[b] The litre actually equals $1{\cdot}000\ 027 \times 10^{-3}$ m³.
[c] The U.S. liquid (not dry) pint. In both U.S. and U.K. measures, two pints = 1 quart.
[d] The U.S. heck is also equal to two U.S. gallons.
[e] The firkin = 9 gallons.

TABLE 15(c) VOLUME (l^3)

Large

Multiply→ by↘ Obtain ↓	Cubic* metre	U.S. fluid ounce	U.K. fluid ounce	U.S. bushel	U.K. bushel	U.S. barrel	U.K. barrel
Cubic metre* (m^3)	1	$2 \cdot 957 \times 10^{-5}$	$2 \cdot 841 \times 10^{-5}$	$3 \cdot 524 \times 10^{-2}$	$3 \cdot 637 \times 10^{-2}$	$1 \cdot 192 \times 10^{-1}$	$1 \cdot 637 \times 10^{-1}$
U.S. fluid ounce[a]	$3 \cdot 378 \times 10^4$	1	$1 \cdot 041$	$1 \cdot 190 \times 10^3$	$1 \cdot 229 \times 10^3$	$4 \cdot 027 \times 10^3$	$5 \cdot 530 \times 10^3$
U.K. fluid ounce[a]	$3 \cdot 521 \times 10^4$	$9 \cdot 615 \times 10^{-1}$	1	$1 \cdot 241 \times 10^3$	$1 \cdot 281 \times 10^3$	$4 \cdot 197 \times 10^3$	$5 \cdot 764 \times 10^3$
U.S. bushel[b]	$2 \cdot 838 \times 10$	$8 \cdot 392 \times 10^{-4}$	$8 \cdot 063 \times 10^{-4}$	1	$1 \cdot 032$	$3 \cdot 281$	$4 \cdot 646$
U.K. bushel[b]	$2 \cdot 747 \times 10$	$8 \cdot 123 \times 10^{-4}$	$7 \cdot 804 \times 10^{-4}$	$9 \cdot 690 \times 10^{-1}$	1	$3 \cdot 274$	$4 \cdot 497$
U.S. barrel[c]	$8 \cdot 403$	$2 \cdot 485 \times 10^{-4}$	$2 \cdot 387 \times 10^{-4}$	$3 \cdot 048 \times 10^{-1}$	$3 \cdot 056 \times 10^{-1}$	1	$1 \cdot 376$
U.K. barrel[c]	$6 \cdot 098$	$1 \cdot 803 \times 10^{-4}$	$1 \cdot 732 \times 10^{-4}$	$2 \cdot 149 \times 10^{-1}$	$2 \cdot 218 \times 10^{-1}$	$7 \cdot 269 \times 10^{-1}$	1

Notes: [a] The drachm $= 1 \cdot 250 \times 10^{-1}$ of an U.K. fluid ounce while the minim $= 2 \cdot 083 \times 10^{-3}$ of a U.K. fluid ounce.
  [b] Dry bushels.
  [c] Dry barrels—the U.S. liquid barrel is equal to $31 \cdot 5$ U.S. gallons, or $1 \cdot 192 \times 10^{-1} m^3$, or 42 U.S. gallons of oil.

TABLE 16 ANGLE

Multiply→ by↘ Obtain ↓	Radian*	Revolution	Degree	Minute	Second
Radian* (rad)	1	6·283	$2·262 \times 10^3$	$1·357 \times 10^5$	$8·143 \times 10^6$
Revolution (rev)	$1·592 \times 10^{-1}$	1	$2·778 \times 10^{-3}$	$4·630 \times 10^{-5}$	$7·692 \times 10^{-7}$
Degree (°)	$4·425 \times 10^{-4}$	$3·600 \times 10^2$	1	$1·667 \times 10^{-2}$	$2·778 \times 10^{-4}$
Minute	$7·353 \times 10^{-6}$	$2·160 \times 10^4$	$6·000 \times 10$	1	$1·667 \times 10^{-2}$
Second	$1·229 \times 10^{-7}$	$1·296 \times 10^6$	$3·600 \times 10^3$	$6·000 \times 10$	1

TABLE 17 TIME (t)

Multiply→ by↘ Obtain ↓	Second*	Minute	Hour	Solar day	Sidereal day	Solar year
Second* (s)	1	$6·000 \times 10$	$3·600 \times 10^2$	$8·640 \times 10^4$	$8·616 \times 10^4$	$3·156 \times 10^7$
Minute (min)	$1·667 \times 10^{-2}$	1	$6·000 \times 10$	$1·440 \times 10^3$	$1·436 \times 10^3$	$5·259 \times 10^5$
Hour (hr)	$2·778 \times 10^{-4}$	$1·667 \times 10^{-2}$	1	$2·400 \times 10$	$2·360 \times 10$	$8·765 \times 10^3$
Solar day	$1·157 \times 10^{-5}$	$6·944 \times 10^{-4}$	$4·167 \times 10^{-2}$	1	$9·833 \times 10^{-1}$	$3·652 \times 10^2$
Sidereal day	$1·161 \times 10^{-5}$	$6·944 \times 10^{-4}$	$4·237 \times 10^2$	$1·017$	1	$3·591 \times 10^2$
Solar year	$3·165 \times 10^{-8}$	$1·901 \times 10^{-6}$	$1·140 \times 10^{-4}$	$2·740 \times 10^{-3}$	$2·786 \times 10^{-3}$	1

TABLE 18 VELOCITY (*u*)

Multiply→ by ↘ Obtain ↓	Metre/ second*	Centi- metre/ second	Kilo- metre/ second	Kilo- metre/ hour	Foot/ second	Mile/ hour	Knot
Metre/second*	1	$1\cdot000 \times 10^{-2}$	$1\cdot000 \times 10^{3}$	$2\cdot778 \times 10^{-1}$	$3\cdot048 \times 10^{-1}$	$4\cdot470 \times 10^{-1}$	$5\cdot155 \times 10^{-1}$
Centimetre/ second	$1\cdot000 \times 10^{2}$	1	$1\cdot000 \times 10^{5}$	$2\cdot778 \times 10$	$3\cdot048 \times 10$	$4\cdot470 \times 10$	$5\cdot155 \times 10$
Kilometre/ second	$1\cdot000 \times 10^{-3}$	$1\cdot000 \times 10^{-5}$	1	$2\cdot778 \times 10^{-4}$	$3\cdot048 \times 10^{-4}$	$4\cdot470 \times 10^{-4}$	$5\cdot155 \times 10^{-4}$
Kilometre/hour	$3\cdot600$	$3\cdot600 \times 10^{-2}$	$3\cdot600 \times 10^{3}$	1	$1\cdot097$	$1\cdot609$	$1\cdot852$
Foot/second	$3\cdot281$	$3\cdot281 \times 10^{-2}$	$3\cdot281 \times 10^{3}$	$9\cdot113 \times 10^{-1}$	1	$1\cdot467$	$1\cdot689$
Mile/hour	$2\cdot237$	$2\cdot237 \times 10^{-2}$	$2\cdot237 \times 10^{3}$	$6\cdot214 \times 10^{-1}$	$6\cdot818 \times 10^{-1}$	1	$1\cdot152$
Knot	$1\cdot943$	$1\cdot943 \times 10^{-2}$	$1\cdot943 \times 10^{3}$	$5\cdot396 \times 10^{-1}$	$5\cdot921 \times 10^{-1}$	$8\cdot684 \times 10^{-1}$	1

TABLE 19(a) MASS (*m*)

Small

Multiply→ by↘ Obtain↓	Kilo-gramme*	Gramme	Grain	Dram	Ounce	Pound	Atomic mass unit
Kilogramme* (kg)	1	$1 \cdot 000 \times 10^{-3}$	$6 \cdot 480 \times 10^{-5}$	$1 \cdot 772 \times 10^{-3}$	$2 \cdot 835 \times 10^{-2}$	$4 \cdot 536 \times 10^{-1}$	$1 \cdot 660 \times 10^{-27}$
Gramme (g)	$1 \cdot 000 \times 10^{3}$	1	$6 \cdot 480 \times 10^{-2}$	$1 \cdot 772$	$2 \cdot 835 \times 10$	$4 \cdot 536 \times 10^{2}$	$1 \cdot 660 \times 10^{-24}$
Grain[a] (gr)	$1 \cdot 543 \times 10^{4}$	$1 \cdot 543 \times 10$	1	$2 \cdot 734 \times 10$	$4 \cdot 375 \times 10^{2}$	$6 \cdot 990 \times 10^{3}$	$2 \cdot 561 \times 10^{-23}$
Dram[b] (dr)	$5 \cdot 644 \times 10^{2}$	$5 \cdot 644 \times 10^{-1}$	$3 \cdot 657 \times 10^{-2}$	1	$1 \cdot 600 \times 10$	$2 \cdot 560 \times 10^{2}$	$9 \cdot 369 \times 10^{-25}$
Ounce[b] (oz)	$3 \cdot 527 \times 10$	$3 \cdot 527 \times 10^{-2}$	$2 \cdot 286 \times 10^{-3}$	$6 \cdot 250 \times 10^{-2}$	1	$1 \cdot 600 \times 10$	$5 \cdot 854 \times 10^{-26}$
Pound[b] (lb)	$2 \cdot 205$	$2 \cdot 205 \times 10^{-3}$	$1 \cdot 429 \times 10^{-4}$	$3 \cdot 906 \times 10^{-3}$	$6 \cdot 250 \times 10^{-2}$	1	$3 \cdot 660 \times 10^{-27}$
Atomic mass unit (amu)	$6 \cdot 024 \times 10^{26}$	$6 \cdot 024 \times 10^{23}$	$3 \cdot 904 \times 10^{19}$	$1 \cdot 607 \times 10^{24}$	$1 \cdot 708 \times 10^{25}$	$2 \cdot 732 \times 10^{26}$	1

Notes: [a] The grain is the mass common to the three Imperial systems of weight, namely avoirdupois, troy and apothecary. Thus

$$1 \text{ grain} = 2 \cdot 083 \times 10^{-3} \text{ oz (troy or apothecary)}$$
$$= 2 \cdot 286 \times 10^{-3} \text{ oz (avoirdupois).}$$

[b] Dram, ounce and pound are all avoirdupois.

TABLE 19(b) MASS (m)

Large

Multiply→ by↘ Obtain ↓	Kilo-gramme*	Metric slug	Tonne	Pound	Slug	Hun-dred-weight	U.S. or U.K. ton
Kilogramme* (kg)	1	9·806	$1·000$ $\times 10^3$	$4·536$ $\times 10^{-1}$	$1·459$ $\times 10$	$5·080$ $\times 10$	$1·016$ $\times 10^3$
Metric slug	$1·021$ $\times 10^{-1}$	1	$1·021$ $\times 10^2$	$4·631$ $\times 10^{-2}$	$1·490$	$5·187$	$1·037$ $\times 10^2$
Tonne (metric ton)	$1·000$ $\times 10^{-3}$	$9·806$ $\times 10^{-3}$	1	$4·536$ $\times 10^{-4}$	$1·459$ $\times 10^{-2}$	$5·080$ $\times 10^{-2}$	$1·016$
Pound (lb) (avoirdupois)	$2·205$	$2·162$ $\times 10$	$2·205$ $\times 10^3$	1	$3·217$ $\times 10$	$1·120$ $\times 10^2$	$2·240$ $\times 10^3$
Slug (geepound)	$6·854$ $\times 10^{-2}$	$6·721$ $\times 10^{-1}$	$6·854$ $\times 10$	$3·109$ $\times 10^{-2}$	1	$3·482$	$6·963$ $\times 10$
Hundredweight[a] (cwt)	$1·968$ $\times 10^{-2}$	$1·930$ $\times 10^{-1}$	$1·968$ $\times 10$	$8·929$ $\times 10^{-3}$	$2·871$ $\times 10^{-1}$	7	$2·000$ $\times 10$
U.S. or U.K. ton[a]	$9·842$ $\times 10^{-4}$	$9·651$ $\times 10^{-3}$	$9·842$ $\times 10^{-1}$	$4·464$ $\times 10^{-4}$	$1·436$ $\times 10^{-2}$	$5·000$ $\times 10^{-2}$	1

Notes: [a] These are long hundredweights and tons.
The so-called short ton = 20 short cwt.
2 000 lb (avoirdupois).

TABLE 20 FORCE (*F*)

Multiply → by ↘ Obtain ↓	Newton*	Dyne	Gramme force	Kilo- gramme force	Poundal	Pound force
Newton* (N)	1	$1 \cdot 000 \times 10^{-5}$	$9 \cdot 807 \times 10^{-3}$	$9 \cdot 807$	$1 \cdot 383 \times 10^{-1}$	$4 \cdot 448$
Dyne	$1 \cdot 000 \times 10^{5}$	1	$9 \cdot 807 \times 10^{2}$	$9 \cdot 807 \times 10^{5}$	$1 \cdot 383 \times 10^{4}$	$4 \cdot 448 \times 10^{5}$
Gramme force	$1 \cdot 020 \times 10^{2}$	$1 \cdot 020 \times 10^{-3}$	1	$1 \cdot 000 \times 10^{3}$	$1 \cdot 410 \times 10$	$4 \cdot 534 \times 10^{2}$
Kilogramme force	$1 \cdot 020 \times 10^{-1}$	$1 \cdot 020 \times 10^{-6}$	$1 \cdot 000 \times 10^{-3}$	1	$1 \cdot 410 \times 10^{-2}$	$4 \cdot 534 \times 10^{-1}$
Poundal	$7 \cdot 233$	$7 \cdot 233 \times 10^{-5}$	$7 \cdot 093 \times 10^{-2}$	$7 \cdot 093 \times 10$	1	$3 \cdot 217 \times 10$
Pound force	$2 \cdot 248 \times 10^{-1}$	$2 \cdot 248 \times 10^{-6}$	$2 \cdot 208 \times 10^{-3}$	$2 \cdot 208$	$3 \cdot 108 \times 10^{-2}$	1

TABLE 21(a) PRESSURE (p)

Small

Multiply→ by↘ Obtain ↓	Newton/ metre²*	Torr	Milli- bar	Pound- force/ ft²	Kg- force/ m²	Poundal/ ft²	Dyne/ cm²
Newton/metre²* (N m⁻²)^{cdef}	1	$1 \cdot 333$ $\times 10^2$	$1 \cdot 000$ $\times 10^2$	$4 \cdot 788$ $\times 10$	$9 \cdot 807$	$1 \cdot 488$	$1 \cdot 000$ $\times 10^{-1}$
Torr (mm Hg)	$7 \cdot 501$ $\times 10^{-3}$	1	$7 \cdot 501$ $\times 10^{-1}$	$3 \cdot 591$ $\times 10^{-1}$	$7 \cdot 356$ $\times 10^{-2}$	$1 \cdot 116$ $\times 10^{-2}$	$7 \cdot 501$ $\times 10^{-4}$
Millibar (mbar)	$1 \cdot 000$ $\times 10^{-2}$	$1 \cdot 333$	1	$4 \cdot 788$ $\times 10^{-1}$	$9 \cdot 807$ $\times 10^{-2}$	$1 \cdot 488$ $\times 10^{-2}$	$1 \cdot 000$ $\times 10^{-3}$
Pound-force/ft² (lb f ft⁻²)^a	$2 \cdot 089$ $\times 10^{-2}$	$2 \cdot 784$	$2 \cdot 089$	1	$2 \cdot 048$ $\times 10^{-1}$	$3 \cdot 108$ $\times 10^{-2}$	$2 \cdot 089$ $\times 10^{-3}$
Kg-force/m² (Kg f m⁻²)^b	$1 \cdot 020$ $\times 10^{-1}$	$1 \cdot 360$ $\times 10$	$1 \cdot 020$ $\times 10$	$4 \cdot 882$	1	$1 \cdot 517$ $\times 10^{-1}$	$1 \cdot 020$ $\times 10^{-2}$
Poundal/ft² (pdl ft⁻²)	$6 \cdot 720$ $\times 10^{-1}$	$8 \cdot 954$ $\times 10$	$6 \cdot 720$ $\times 10$	$3 \cdot 217$ $\times 10$	$6 \cdot 588$	1	$6 \cdot 720$ $\times 10^{-2}$
Dyne/cm² (dyn cm⁻²)	$1 \cdot 000$ $\times 10$	$1 \cdot 333$ $\times 10^3$	$1 \cdot 000$ $\times 10^3$	$4 \cdot 788$ $\times 10^2$	$9 \cdot 807$ $\times 10$	$1 \cdot 488$ $\times 10$	1

Notes: ^{a} "Pound-force/ft²" is frequently abbreviated to "Pound/ft²".
  ^{b} "Kg-force/m²" is frequently abbreviated to "Kg/m²".
  ^{c} 1 pieze (pz) $= 10^3$N m⁻².
  ^{d} 1 bar $= 10^5$N m⁻².
  ^{e} 1 ton-force/in² $= 1 \cdot 544 \times 10^7$N m⁻².
  ^{f} "Pascal" see page 15 Section 4.

TABLE 21(b) PRESSURE (p)

Large

Multiply→ by↘ Obtain ↓	Newton/ m²*	Atmo- sphere	Kg- force/ cm²	Pound force/ in² (psi)	Inches Hg	Inches H₂O	Mm Hg
Newton/ metre²*ᵉ (N m⁻²)ᶠ	1	$1 \cdot 013 \times 10^5$	$9 \cdot 804 \times 10^4$	$6 \cdot 895 \times 10^3$	$3 \cdot 386 \times 10^3$	$2 \cdot 491 \times 10^2$	$1 \cdot 333 \times 10^2$
Atmosphere (atm)	$9 \cdot 869 \times 10^{-6}$	1	$9 \cdot 678 \times 10^{-1}$	$6 \cdot 804 \times 10^{-2}$	$3 \cdot 342 \times 10^{-2}$	$2 \cdot 458 \times 10^{-3}$	$1 \cdot 316 \times 10^{-3}$
Kg-force/cm² (Kg f cm⁻²)ᶜ	$1 \cdot 020 \times 10^{-5}$	$1 \cdot 033$	1	$7 \cdot 031 \times 10^{-2}$	$3 \cdot 453 \times 10^{-2}$	$2 \cdot 539 \times 10^{-3}$	$1 \cdot 359 \times 10^{-3}$
Pound-force/in² (lb g in⁻² or psi)ᵈ	$1 \cdot 450 \times 10^{-4}$	$1 \cdot 470 \times 10$	$1 \cdot 422 \times 10$	1	$4 \cdot 912 \times 10^{-1}$	$3 \cdot 613 \times 10^{-2}$	$1 \cdot 934 \times 10^{-2}$
Inches Hgᵃ	$2 \cdot 953 \times 10^{-4}$	$2 \cdot 992 \times 10$	$2 \cdot 896 \times 10$	$2 \cdot 036$	1	$7 \cdot 355 \times 10^{-2}$	$3 \cdot 937 \times 10^{-2}$
Inches H₂Oᵇ	$4 \cdot 015 \times 10^{-3}$	$4 \cdot 068 \times 10^2$	$3 \cdot 939 \times 10^2$	$2 \cdot 768 \times 10$	$1 \cdot 360 \times 10$	1	$5 \cdot 354 \times 10^{-1}$
Mm Hgᵃ (torr)	$7 \cdot 501 \times 10^{-3}$	$7 \cdot 600 \times 10^2$	$7 \cdot 355 \times 10^2$	$5 \cdot 171 \times 10$	$2 \cdot 540 \times 10$	$1 \cdot 868$	1

Notes: ᵃ "Inches Hg" and "mm Hg" refers to Hg at 0°C or 273K.
ᵇ "Inches H₂O" refers to water at 4°C or 277K.
ᶜ "Kg force cm⁻²" is frequently abbreviated to "Kg cm⁻²".
ᵈ "Pound force in⁻²" is frequently abbreviated to "pound in⁻²" or "psi".
ᵉ "Newton per square metre".
ᶠ "Pascal" see page 15 Section 4.

TABLE 22(a) ENERGY (*E*)

Part 1

Multiply→ by↘ Obtain ↓	Joule*	Kilo-watt-hour	Kilo-calorie	British ther-mal unit	Foot-pound	Foot-poundal	Litre-atmo-sphere	Erg
Joule (J)*a	1	$3 \cdot 600 \times 10^6$	$4 \cdot 186 \times 10^3$	$1 \cdot 055 \times 10^3$	$1 \cdot 356$	$4 \cdot 214 \times 10^{-2}$	$1 \cdot 013 \times 10^2$	$1 \cdot 000 \times 10^{-7}$
Kilowatt-hour (kWh)	$2 \cdot 778 \times 10^{-7}$	1	$1 \cdot 163 \times 10^{-3}$	$2 \cdot 930 \times 10^{-4}$	$3 \cdot 766 \times 10^{-7}$	$1 \cdot 171 \times 10^{-8}$	$2 \cdot 815 \times 10^{-5}$	$2 \cdot 778 \times 10^{-14}$
Kilocalorie (kcal)	$2 \cdot 389 \times 10^{-4}$	$8 \cdot 600 \times 10^2$	1	$2 \cdot 520 \times 10^{-1}$	$3 \cdot 239 \times 10^{-4}$	$1 \cdot 007 \times 10^{-5}$	$2 \cdot 421 \times 10^{-2}$	$2 \cdot 389 \times 10^{-11}$
British thermal unit (BTU)	$9 \cdot 480 \times 10^{-4}$	$3 \cdot 412 \times 10^3$	$3 \cdot 969$	1	$1 \cdot 285 \times 10^{-3}$	$3 \cdot 994 \times 10^{-5}$	$9 \cdot 604 \times 10^{-2}$	$9 \cdot 480 \times 10^{-11}$
Foot-pound (ft-lb)	$7 \cdot 377 \times 10^{-1}$	$2 \cdot 655 \times 10^6$	$3 \cdot 086 \times 10^3$	$7 \cdot 783 \times 10^2$	1	$3 \cdot 108 \times 10^{-2}$	$7 \cdot 474 \times 10$	$7 \cdot 367 \times 10^{-8}$
Foot-poundal (ft-pdl)	$2 \cdot 373 \times 10$	$8 \cdot 539 \times 10^7$	$9 \cdot 929 \times 10^4$	$2 \cdot 502 \times 10^4$	$3 \cdot 217 \times 10$	1	$2 \cdot 405 \times 10^3$	$2 \cdot 373 \times 10^{-6}$
Litre-atmo-sphere (l-atm)	$9 \cdot 871 \times 10^{-3}$	$3 \cdot 550 \times 10^4$	$4 \cdot 129 \times 10$	$1 \cdot 041 \times 10$	$1 \cdot 338 \times 10^{-2}$	$4 \cdot 159 \times 10^{-4}$	1	$9 \cdot 869 \times 10^{-10}$
Erg	$1 \cdot 000 \times 10^7$	$3 \cdot 600 \times 10^{13}$	$4 \cdot 186 \times 10^{10}$	$1 \cdot 055 \times 10^{10}$	$1 \cdot 356 \times 10^7$	$4 \cdot 214 \times 10^5$	$1 \cdot 013 \times 10^9$	1

Note: [a] 1 Thermie $= 4 \cdot 186 \mathrm{MJ} = 4 \cdot 186 \times 10^6 \, \mathrm{J}$.

TABLE 22(b) ENERGY (E)

Part 2

Multiply→ by↘ Obtain ↓	Joule*	Kilo-gramme	Calorie	Erg	Atomic mass unit	Mev	Elec-tron volt	Centi-metre
Joule (J)*	1	$8 \cdot 987$ $\times 10^{16}$	$4 \cdot 187$	$1 \cdot 000$ $\times 10^{-7}$	$1 \cdot 492$ $\times 10^{-10}$	$1 \cdot 602$ $\times 10^{-13}$	$1 \cdot 602$ $\times 10^{-19}$	$1 \cdot 986$ $\times 10^{-23}$
Kilogramme (kg)[a]	$1 \cdot 113$ $\times 10^{-17}$	1	$4 \cdot 659$ $\times 10^{-17}$	$1 \cdot 113$ $\times 10^{-24}$	$1 \cdot 660$ $\times 10^{-27}$	$1 \cdot 783$ $\times 10^{-30}$	$1 \cdot 783$ $\times 10^{-36}$	$2 \cdot 210$ $\times 10^{-40}$
Calorie (cal)[bcd]	$2 \cdot 389$ $\times 10^{-1}$	$2 \cdot 147$ $\times 10^{16}$	1	$2 \cdot 389$ $\times 10^{-8}$	$3 \cdot 564$ $\times 10^{-11}$	$3 \cdot 827$ $\times 10^{-14}$	$3 \cdot 827$ $\times 10^{-20}$	$4 \cdot 745$ $\times 10^{-24}$
Erg	$1 \cdot 000$ $\times 10^{7}$	$8 \cdot 987$ $\times 10^{23}$	$4 \cdot 187$ $\times 10^{7}$	1	$1 \cdot 492$ $\times 10^{-3}$	$1 \cdot 602$ $\times 10^{-6}$	$1 \cdot 602$ $\times 10^{-12}$	$1 \cdot 986$ $\times 10^{-16}$
Atomic mass unit (amu)	$6 \cdot 701$ $\times 10^{9}$	$6 \cdot 025$ $\times 10^{26}$	$2 \cdot 807$ $\times 10^{10}$	$6 \cdot 701$ $\times 10^{2}$	1	$1 \cdot 074$ $\times 10^{-3}$	$1 \cdot 074$ $\times 10^{-9}$	$1 \cdot 331$ $\times 10^{-13}$
Mev	$6 \cdot 242$ $\times 10^{12}$	$5 \cdot 610$ $\times 10^{29}$	$2 \cdot 613$ $\times 10^{13}$	$6 \cdot 242$ $\times 10^{5}$	$9 \cdot 315$ $\times 10^{2}$	1	$1 \cdot 000$ $\times 10^{-6}$	$1 \cdot 240$ $\times 10^{-10}$
Electron volt (ev)	$6 \cdot 242$ $\times 10^{18}$	$5 \cdot 610$ $\times 10^{35}$	$2 \cdot 613$ $\times 10^{19}$	$6 \cdot 242$ $\times 10^{11}$	$9 \cdot 315$ $\times 10^{8}$	$1 \cdot 000$ $\times 10^{6}$	1	$1 \cdot 240$ $\times 10^{-4}$
Centimetre^{-1} (cm^{-1})	$5 \cdot 035$ $\times 10^{22}$	$4 \cdot 524$ $\times 10^{39}$	$2 \cdot 108$ $\times 10^{23}$	$5 \cdot 035$ $\times 10^{15}$	$7 \cdot 513$ $\times 10^{12}$	$8 \cdot 066$ $\times 10^{9}$	$8 \cdot 066$ $\times 10^{3}$	1

Notes: [a] The conversion factors for kilogrammes and atomic mass used were obtained by use of the mass-energy relationship $E = mc^2$.

[b] 1 15°C calorie $= 4 \cdot 1855$J.

[c] 1 International Table calorie $= 4 \cdot 1868$J.

[d] 1 thermochemical calorie $= 4 \cdot 1840$J.

TABLE 23(a) POWER (*P*)

Part 1

Multiply→ by↘ Obtain ↓	Watt*	Erg/ second	Kcal minute	Foot- pound/ second	H.P. (British)	BTU min^{-1}	Kilo- watt
Watt (W)* or Joule s^{-1} (J s^{-1})	1	$1 \cdot 000$ $\times 10^{-7}$	$6 \cdot 977$ $\times 10$	$1 \cdot 356$	$7 \cdot 457$ $\times 10^2$	$1 \cdot 758$ $\times 10$	$1 \cdot 000$ $\times 10^3$
Erg/second	$1 \cdot 000$ $\times 10^7$	1	$6 \cdot 977$ $\times 10^8$	$1 \cdot 356$ $\times 10^7$	$7 \cdot 457$ $\times 10^9$	$1 \cdot 758$ $\times 10^8$	$1 \cdot 000$ $\times 10^{10}$
Kcal/minute	$1 \cdot 433$ $\times 10^{-2}$	$1 \cdot 433$ $\times 10^{-9}$	1	$1 \cdot 943$ $\times 10^{-2}$	$1 \cdot 069$ $\times 10$	$2 \cdot 520$ $\times 10^{-1}$	$1 \cdot 433$ $\times 10$
Foot-pound/ second^a	$7 \cdot 376$ $\times 10^{-1}$	$7 \cdot 376$ $\times 10^{-8}$	$5 \cdot 144$ $\times 10$	1	$5 \cdot 500$ $\times 10^2$	$1 \cdot 297$ $\times 10$	$7 \cdot 376$ $\times 10^2$
HP (British)	$1 \cdot 341$ $\times 10^{-3}$	$1 \cdot 341$ $\times 10^{-10}$	$9 \cdot 355$ $\times 10^{-2}$	$1 \cdot 818$ $\times 10^{-3}$	1	$2 \cdot 357$ $\times 10^{-2}$	$1 \cdot 341$
BTU/min	$5 \cdot 689$ $\times 10^{-2}$	$5 \cdot 689$ $\times 10^{-9}$	$3 \cdot 969$	$7 \cdot 712$ $\times 10^{-2}$	$4 \cdot 241$ $\times 10$	1	$5 \cdot 689$ $\times 10$
Kilowatt (kw)	$1 \cdot 000$ $\times 10^{-3}$	$1 \cdot 000$ $\times 10^{-10}$	$6 \cdot 977$ $\times 10^{-2}$	$1 \cdot 356$ $\times 10^{-3}$	$7 \cdot 457$ $\times 10^{-1}$	$1 \cdot 758$ $\times 10^{-2}$	1

Note: ^{a} Foot-pound/second is an abbreviation for foot-pound-force/second (ft lb f s^{-1}).

TABLE 23(b) POWER (*P*)

Part 2

Multiply→ by↘ Obtain↓	Watt* or J s^{-1}	Calorie/ second	Kilo- gramme metre/ second	Foot- pound/ second	Foot- poun- dal/ second	BTU/ hour	HP (Brit- ish)	HP (Met- ric)
Watt (W)* or Joule s^{-1} (J s^{-1})	1	4·187	9·807	1·356	4·214 $\times 10^{-2}$	2·931 $\times 10^{-1}$	7·457 $\times 10^{2}$	7·355 $\times 10^{2}$
Calorie/second (cal s^{-1})	2·388 $\times 10^{-1}$	1	2·343	3·239 $\times 10^{-1}$	1·007 $\times 10^{-2}$	6·999 $\times 10^{-2}$	1·782 $\times 10^{2}$	1·757 $\times 10^{2}$
Kilogramme- metre/second (kg m s^{-1})	1·020 $\times 10^{-1}$	4·268 $\times 10^{-1}$	1	1·383 $\times 10^{-1}$	4·296 $\times 10^{-3}$	2·987 $\times 10^{-2}$	7·604 $\times 10$	7·500 $\times 10$
Foot-pound/ second[a]	7·376 $\times 10^{-1}$	3·087	7·233	1	3·108 $\times 10^{-2}$	2·161 $\times 10^{-1}$	5·500 $\times 10^{2}$	5·425 $\times 10^{2}$
Foot-poundal/ second (Ft-pdl s^{-1})	2·373 $\times 10$	9·933 $\times 10$	2·328 $\times 10^{2}$	3·218 $\times 10$	1	6·955	1·770 $\times 10^{4}$	1·746 $\times 10^{4}$
BTU/hour	3·413	1·429 $\times 10$	3·347 $\times 10$	4·628	1·438 $\times 10^{-1}$	1	2·545 $\times 10^{3}$	2·511 $\times 10^{3}$
HP (British)	1·341 $\times 10^{-3}$	5·613 $\times 10^{-3}$	1·315 $\times 10^{-2}$	1·818 $\times 10^{-3}$	5·649 $\times 10^{-5}$	3·928 $\times 10^{-4}$	1	9·863 $\times 10^{-1}$
HP (Metric)	1·360 $\times 10^{-3}$	5·694 $\times 10^{-3}$	1·333 $\times 10^{-2}$	1·845 $\times 10^{-3}$	5·728 $\times 10^{-5}$	3·982 $\times 10^{-4}$	1·014	1

Note: [a] Foot-pound/second is an abbreviation for foot-pound force second^{-1} (ft lb f s^{-1}).

TABLE 24 ELECTRIC CHARGE (Q)

Multiply→ by↘ Obtain ↓	Coulomb (C)*	Ampere-hour	Ab-coulomb	Faraday	Stat-coulomb	Franklin
Coulomb (C)*[a] (ampere-second)	1	$3 \cdot 600 \times 10^3$	$1 \cdot 000 \times 10$	$9 \cdot 649 \times 10^4$	$3 \cdot 336 \times 10^{-10}$	$3 \cdot 336 \times 10^{-10}$
Ampere-hour	$2 \cdot 778 \times 10^{-4}$	1	$2 \cdot 778 \times 10^{-3}$	$2 \cdot 681 \times 10$	$9 \cdot 266 \times 10^{-14}$	$9 \cdot 266 \times 10^{-14}$
Abcoulomb	$1 \cdot 000 \times 10^{-1}$	$3 \cdot 600 \times 10^2$	1	$9 \cdot 649 \times 10^3$	$3 \cdot 336 \times 10^{-11}$	$3 \cdot 336 \times 10^{-11}$
Faraday	$1 \cdot 036 \times 10^{-5}$	$3 \cdot 730 \times 10^{-2}$	$1 \cdot 036 \times 10^{-4}$	1	$3 \cdot 457 \times 10^{-15}$	$3 \cdot 457 \times 10^{-15}$
Statcoulomb	$2 \cdot 998 \times 10^9$	$1 \cdot 079 \times 10^{13}$	$2 \cdot 998 \times 10^{10}$	$2 \cdot 893 \times 10^{14}$	1	$1 \cdot 000$
Franklin (Fr)	$2 \cdot 998 \times 10^9$	$1 \cdot 079 \times 10^{13}$	$2 \cdot 998 \times 10^{10}$	$2 \cdot 893 \times 10^{14}$	$1 \cdot 000$	1

Note: [a] 1 international coulomb = 0·999 835 absolute coulomb.

TABLE 25 ELECTRIC CURRENT (I)

Multiply→ by↘ Obtain ↓	Ampere (A)*	Abampere	Biot	Stat-ampere
Ampere (A)*[a]	1	$1 \cdot 000 \times 10$	$1 \cdot 000 \times 10$	$3 \cdot 336 \times 10^{-10}$
Abampere (ab A)	$1 \cdot 000 \times 10^{-1}$	1	$1 \cdot 000$	$3 \cdot 336 \times 10^{-11}$
Biot (Bi)	$1 \cdot 000 \times 10^{-1}$	$1 \cdot 000$	1	$3 \cdot 336 \times 10^{-11}$
Statampere	$2 \cdot 998 \times 10^9$	$2 \cdot 998 \times 10^{10}$	$2 \cdot 998 \times 10^{10}$	1

Note: [a] 1 international ampere = 0·999 835 absolute ampere.

TABLE 26 ELECTRIC POTENTIAL (*V*)

Multiply→ by↘ Obtain ↓	Volt (V)*	Statvolt	Millivolt	Microvolt	Abvolt
Volt (V)*ᵃ	1	$2 \cdot 998$ $\times 10^2$	$1 \cdot 000$ $\times 10^{-3}$	$1 \cdot 000$ $\times 10^{-6}$	$1 \cdot 000$ $\times 10^{-8}$
Statvolt	$3 \cdot 336$ $\times 10^{-3}$	1	$3 \cdot 336$ $\times 10^{-6}$	$3 \cdot 336$ $\times 10^{-9}$	$3 \cdot 336$ $\times 10^{-11}$
Millivolt (mV)	$1 \cdot 000$ $\times 10^3$	$2 \cdot 998$ $\times 10^5$	1	$1 \cdot 000$ $\times 10^{-3}$	$1 \cdot 000$ $\times 10^{-5}$
Microvolt (µV)	$1 \cdot 000$ $\times 10^6$	$2 \cdot 998$ $\times 10^8$	$1 \cdot 000$ $\times 10^3$	1	$1 \cdot 000$ $\times 10^{-2}$
Abvolt (abV)	$1 \cdot 000$ $\times 10^8$	$2 \cdot 998$ $\times 10^{10}$	$1 \cdot 000$ $\times 10^5$	$1 \cdot 000$ $\times 10^2$	1

Note: ᵃ 1 international volt = $1 \cdot 000$ 330 absolute volts.

TABLE 27 ELECTRIC FIELD STRENGTH (*E*)

Multiply→ by↘ Obtain ↓	Volt m^{-1}*	Statvolt cm^{-1}	Volt cm^{-1}	Abvolt cm^{-1}	Volt in^{-1}
Volt metre⁻¹ (V m⁻¹)*	1	$2 \cdot 998$ $\times 10^4$	$1 \cdot 000$ $\times 10^2$	$1 \cdot 000$ $\times 10^{-6}$	$3 \cdot 937$ $\times 10$
Statvolt cm⁻¹	$3 \cdot 336$ $\times 10^{-5}$	1	$3 \cdot 336$ $\times 10^{-3}$	$3 \cdot 336$ $\times 10^{-11}$	$1 \cdot 313$ $\times 10^{-3}$
Volt cm⁻¹ (V cm⁻¹)	$1 \cdot 000$ $\times 10^{-2}$	$2 \cdot 998$ $\times 10^2$	1	$1 \cdot 000$ $\times 10^{-8}$	$3 \cdot 937$ $\times 10^{-1}$
Abvolt cm⁻¹ (ab V cm⁻¹)	$1 \cdot 000$ $\times 10^6$	$2 \cdot 998$ $\times 10^{10}$	$1 \cdot 000$ $\times 10^8$	1	$3 \cdot 937$ $\times 10^7$
Volt inch⁻¹ (V in⁻¹)	$2 \cdot 540$ $\times 10^{-2}$	$7 \cdot 615$ $\times 10^2$	$2 \cdot 540$	$2 \cdot 540$ $\times 10^{-8}$	1

TABLE 28 ELECTRIC RESISTANCE (R)

Multiply→ by↘ Obtain ↓	Ohm*	Gigohm	Statohm	Megohm	Abohm
Ohm (Ω)*[a]	1	$1 \cdot 000 \times 10^9$	$8 \cdot 988 \times 10^{11}$	$1 \cdot 000 \times 10^6$	$1 \cdot 000 \times 10^{-9}$
Gigohm (GΩ)	$1 \cdot 000 \times 10^{-9}$	1	$8 \cdot 988 \times 10^2$	$1 \cdot 000 \times 10^{-3}$	$1 \cdot 000 \times 10^{-1}$
Statohm	$1 \cdot 113 \times 10^{-12}$	$1 \cdot 113 \times 10^{-3}$	1	$1 \cdot 113 \times 10^{-6}$	$1 \cdot 113 \times 10^{-2}$
Megohm (MΩ)	$1 \cdot 000 \times 10^{-6}$	$1 \cdot 000 \times 10^3$	$8 \cdot 988 \times 10^5$	1	$1 \cdot 000 \times 10^{-1}$
Abohm (abΩ)	$1 \cdot 000 \times 10^9$	$1 \cdot 000 \times 10^{18}$	$8 \cdot 988 \times 10^{20}$	$1 \cdot 000 \times 10^{15}$	1

Note: [a] 1 international ohm = $1 \cdot 000\,495$ absolute ohms.

TABLE 29 ELECTRIC RESISTIVITY (ρ)
(Formerly called Specific Resistance)

Multiply→ by↘ Obtain ↓	Ohm-m*	Statohm cm	Ohm-in	Ohm-cm	Abohm- cm
Ohm-m (Ω m)*	1	$8 \cdot 988 \times 10^9$	$2 \cdot 540 \times 10^{-2}$	$1 \cdot 000 \times 10^{-2}$	$1 \cdot 000 \times 10^{-1}$
Statohm-cm	$1 \cdot 113 \times 10^{-10}$	1	$2 \cdot 827 \times 10^{-12}$	$1 \cdot 113 \times 10^{-12}$	$1 \cdot 113 \times 10^{-2}$
Ohm-inches (Ω in)	$3 \cdot 937 \times 10$	$3 \cdot 539 \times 10^{11}$	1	$3 \cdot 937 \times 10^{-1}$	$3 \cdot 937 \times 10^{-1}$
Ohm-cm (Ω cm)	$1 \cdot 000 \times 10^2$	$8 \cdot 988 \times 10^{11}$	$2 \cdot 540$	1	$1 \cdot 000 \times 10^{-9}$
Abohm-cm (ab Ω cm)	$1 \cdot 000 \times 10^{11}$	$8 \cdot 988 \times 10^{20}$	$2 \cdot 540 \times 10^9$	$1 \cdot 000 \times 10^9$	1

TABLE 30 ELECTRIC CONDUCTANCE (*G*)

Multiply→ by↘ Obtain ↓	Siemens (S)*	Abmho	Mho	Micro-siemens	Statmho
Siemens (S)* or Ohm^{-1} (Ω$^-$)	1	1.000×10^9	1.000	1.000×10^{-6}	1.113×10^{-12}
Abmho	1.000×10^{-9}	1	1.000×10^{-9}	1.000×10^{-15}	1.113×10^{-21}
Mho (Ω^{-1})[a]	1.000	1.000×10^9	1	1.000×10^{-6}	1.113×10^{-12}
Microsiemens (μS)[b] or Megohm^{-1}	1.000×10^6	1.000×10^{15}	1.000×10^6	1	1.113×10^{-6}
Statmho	8.988×10^{11}	8.988×10^{20}	8.988×10^{11}	8.922×10^5	1

Notes: [a] International mho = 0.999 505 mho.
[b] Microsiemens or megohm^{-1} have been called micromho or gemmho.

TABLE 31 CAPACITANCE (*C*)

Multiply→ by↘ Obtain ↓	Farad*	Abfarad	Micro-farad	Statfarad	Picofarad
Farad (F)*[a]	1	1.000×10^9	1.000×10^{-6}	1.113×10^{-12}	1.000×10^{-12}
Abfarad (abF)	1.000×10^{-9}	1	1.000×10^{-15}	1.113×10^{-21}	1.000×10^{-21}
Microfarad (μF)	1.000×10^6	1.000×10^{15}	1	1.113×10^{-6}	1.000×10^{-6}
Statfarad	8.988×10^{11}	8.988×10^{20}	8.988×10^5	1	8.988×10^{-1}
Picofarad (pF) or "puff"	1.000×10^{12}	1.000×10^{21}	1.000×10^6	1.113	1

Note: [a] 1 international farad = 0.999 505 absolute farads.

TABLE 32 INDUCTANCE (*L*)

Multiply→ by↘ Obtain ↓	Henry*	Stathenry	Milli-henry	Abhenry	Pico-henry
Henry (H)*[a]	1	$8 \cdot 988 \times 10^{11}$	$1 \cdot 000 \times 10^{-3}$	$1 \cdot 000 \times 10^{-9}$	$1 \cdot 000 \times 10^{-12}$
Stathenry	$1 \cdot 113 \times 10^{-12}$	1	$1 \cdot 113 \times 10^{-15}$	$1 \cdot 113 \times 10^{-21}$	$1 \cdot 113 \times 10^{-24}$
Millihenry (mh)	$1 \cdot 000 \times 10^{3}$	$8 \cdot 988 \times 10^{14}$	1	$1 \cdot 000 \times 10^{-6}$	$1 \cdot 000 \times 10^{-9}$
Abhenry (abH)	$1 \cdot 000 \times 10^{9}$	$8 \cdot 988 \times 10^{20}$	$1 \cdot 000 \times 10^{6}$	1	$1 \cdot 000 \times 10^{-3}$
Picohenry (pH)	$1 \cdot 000 \times 10^{12}$	$8 \cdot 988 \times 10^{23}$	$1 \cdot 000 \times 10^{9}$	$1 \cdot 000 \times 10^{3}$	1

Note: [a] 1 international henry = $1 \cdot 000\,495$ absolute henry.
The mic (= 10^{-6} henry) was used by the Royal Navy from 1920 to 1938.

TABLE 33 MAGNETIC FLUX (ϕ)

Multiply→ by↘ Obtain ↓	Weber*	Maxwell	Line	Kiloline
Weber (Wb)*	1	$1 \cdot 000 \times 10^{-8}$	$1 \cdot 000 \times 10^{-8}$	$1 \cdot 000 \times 10^{-5}$
Maxwell (Mx)[a]	$1 \cdot 000 \times 10^{8}$	1	$1 \cdot 000$	$1 \cdot 000 \times 10^{3}$
Line	$1 \cdot 000 \times 10^{8}$	$1 \cdot 000$	1	$1 \cdot 000 \times 10^{3}$
Kiloline	$1 \cdot 000 \times 10^{5}$	$1 \cdot 000 \times 10^{-3}$	$1 \cdot 000 \times 10^{-3}$	1

Note: [a] The promaxwell (= 10^{8} maxwell) was used for a short time after 1930 but has been replaced by the weber.

TABLE 34 MAGNETIC INDUCTION, MAGNETIC FLUX DENSITY (B)

Multiply→ by↘ Obtain ↓	Tesla*	Gauss	Maxwell/ cm^2	Gamma
Tesla (T)* $(Wb\ m^{-2})$	1	$1 \cdot 000 \times 10^{-4}$	$1 \cdot 000 \times 10^{-4}$	$1 \cdot 000 \times 10^{-9}$
Gauss (Gs or G) $(line\ cm^{-2})$	$1 \cdot 000 \times 10^4$	1	$1 \cdot 000$	$1 \cdot 000 \times 10^{-5}$
Maxwell/centimetre^2 $(Mx\ cm^{-2})$	$1 \cdot 000 \times 10^4$	$1 \cdot 000$	1	$1 \cdot 000 \times 10^{-5}$
Gamma[a]	$1 \cdot 000 \times 10^9$	$1 \cdot 000 \times 10^5$	$1 \cdot 000 \times 10^5$	1

Note: [a] The gamma has been used in geophysics since the beginning of the twentieth century.

TABLE 35 MAGNETIC FIELD STRENGTH (H)

Multiply→ by↘ Obtain ↓	Ampere m^{-1}*	Biot cm^{-1}	Abamp cm^{-1}	Oersted	Gilbert cm^{-1}	Ampere in^{-1}
Ampere/metre (Am^{-1})*[ab]	1	$1 \cdot 000 \times 10^3$	$1 \cdot 000 \times 10^3$	$7 \cdot 958 \times 10$	$7 \cdot 958 \times 10$	$3 \cdot 937 \times 10$
Biot/centimetre $(Bi\ cm^{-1})$	$1 \cdot 000 \times 10^{-3}$	1	$1 \cdot 000$	$7 \cdot 958 \times 10^{-2}$	$7 \cdot 958 \times 10^{-2}$	$3 \cdot 937 \times 10^{-2}$
Abampere/centimetre $(abA\ cm^{-1})$	$1 \cdot 000 \times 10^{-3}$	$1 \cdot 000$	1	$7 \cdot 958 \times 10^{-2}$	$7 \cdot 958 \times 10^{-2}$	$3 \cdot 937 \times 10^{-2}$
Oersted (oe)	$1 \cdot 257 \times 10^{-2}$	$1 \cdot 257 \times 10$	$1 \cdot 257 \times 10$	1	$1 \cdot 000$	$4 \cdot 949 \times 10^{-1}$
Gilbert/centimetre $(Gb\ cm^{-1})$	$1 \cdot 257 \times 10^{-2}$	$1 \cdot 257 \times 10$	$1 \cdot 257 \times 10$	$1 \cdot 000$	1	$4 \cdot 949 \times 10^{-1}$
Ampere/inch $(A\ in^{-1})$	$2 \cdot 539 \times 10^{-2}$	$2 \cdot 539 \times 10$	$2 \cdot 539 \times 10$	$2 \cdot 021$	$2 \cdot 021$	1

Notes: [a] The term "ampere-turn" 'has sometimes been used in this context in place of "ampere".
[b] The term "praoersted" has sometimes been used as a unit equal to 4π ampere-turns metre^{-1}.

TABLE 36 MAGNETOMOTIVE FORCE (F_m)

Multiply→ by↘ Obtain ↓	Ampere*	Gilbert	Abampere
Ampere (A)*[ab]	1	$7 \cdot 958$ $\times 10^{-1}$	$1 \cdot 000$ $\times 10$
Gilbert (Gb)	$1 \cdot 257$	1	$1 \cdot 257$ $\times 10$
Abampere (abA)	$1 \cdot 000$ $\times 10^{-1}$	$7 \cdot 958$ $\times 10^{-2}$	1

Notes: [a] The term "ampere-turn" has sometimes been used in this context in place of "ampere".
[b] The term "pragilbert" has sometimes been used as a unit equal to 4π ampere-turns.

TABLE 37 ILLUMINATION

Multiply→ by↘ Obtain ↓	Lux*	Foot-candle	Lumen/ cm^2	Phot
Lux (lx)* (lm m^{-2})	1	$1 \cdot 076$ $\times 10$	$1 \cdot 000$ $\times 10^4$	$1 \cdot 000$ $\times 10^4$
Foot-candle (fc)	$9 \cdot 290$ $\times 10^{-2}$	1	$9 \cdot 290$ $\times 10^2$	$9 \cdot 290$ $\times 10^2$
Lumen/centimetre^2	$1 \cdot 000$ $\times 10^{-4}$	$1 \cdot 076$ $\times 10^{-3}$	1	$1 \cdot 000$
Phot	$1 \cdot 000$ $\times 10^{-4}$	$1 \cdot 076$ $\times 10^{-3}$	$1 \cdot 000$	1

Note: Nox was used by Germany during World War II as a unit. It is equivalent to 10^{-3} lux.

TABLE 38 LUMINANCE

Multiply→ by↘ Obtain ↓	Candela/m²* (nit)	Candela/cm² (stilb)	Lambert	Candela/ft²	Foot-lambert	Apostilb
Candela/metre² (cd m^{-2})* ᵃ or nit (nt)	1	$1 \cdot 000 \times 10^4$	$3 \cdot 183 \times 10^3$	$1 \cdot 076 \times 10$	$3 \cdot 426$	$3 \cdot 183 \times 10^{-1}$
Candela/centimetre² (cd cm^{-2}) or stilb (sb)	$1 \cdot 000 \times 10^{-4}$	1	$3 \cdot 183 \times 10^{-1}$	$1 \cdot 076 \times 10^{-3}$	$3 \cdot 426 \times 10^{-4}$	$3 \cdot 183 \times 10^{-5}$
Lambert (L) or lumen cm^{-2} (lm cm^{-2})	$3 \cdot 142 \times 10^{-4}$	$3 \cdot 142$	1	$3 \cdot 381 \times 10^{-3}$	$1 \cdot 076 \times 10^{-3}$	$1 \cdot 000 \times 10^{-4}$
Candela/foot² (cd ft^{-2})	$9 \cdot 290 \times 10^{-2}$	$9 \cdot 290 \times 10^2$	$2 \cdot 957 \times 10^2$	1	$3 \cdot 183 \times 10^{-1}$	$2 \cdot 957 \times 10^{-2}$
Foot-lambert (ft-L) or equivalent foot candle	$2 \cdot 919 \times 10^{-1}$	$2 \cdot 919 \times 10^3$	$9 \cdot 290 \times 10^2$	$3 \cdot 142$	1	$9 \cdot 290 \times 10^{-2}$
Apostilb (asb)ᵇ or lumen metre^{-2} (lm m^{-2})	$3 \cdot 142$	$3 \cdot 142 \times 10^4$	$1 \cdot 000 \times 10^4$	$3 \cdot 381 \times 10$	$1 \cdot 076 \times 10$	1

Notes: [a] Luminous intensity of candela $= 98 \cdot 1\%$ that of international candle.
[b] The skot was used by Germany for a short period. It is equivalent to 10^{-3} apostilb or 10^{-3} lumen per square metre.

Appendixes

APPENDIX I

Electrical and Magnetic Units

The study of electrostatics and magnetostatics can be developed along mutually similar lines to give two sets of well-established units, both based on the CGS system. Electrostatic units (ESU) and electromagnetic units (EMU) are the simplest units for use in the study of electrostatics and magnetostatics. However, difficulties arise in the study of electrodynamics when quantities expressed in ESU are directly related to quantities expressed in EMU, as in the case where a force is experienced by a moving magnetic charge in the presence of an electric field.

When an attempt is made to use only mechanical dimensions for both electrostatics and magnetostatics, it is found that the electric and magnetic quantities have conflicting dimensions, depending on whether the dimensions were derived from EMU or ESU formulae (see Table 39).

TABLE 39 CURRENT AND CHARGE DIMENSIONS IN ESU AND EMU

Quantity	System dimensions	
	ESU	EMU
Electric current	$(\text{length})^{\frac{3}{2}} (\text{mass})^{\frac{1}{2}} (\text{time})^{-2}$	$(\text{length})^{\frac{1}{2}} (\text{mass})^{\frac{1}{2}} (\text{time})^{-1}$
Electric charge	$(\text{length})^{\frac{3}{2}} (\text{mass})^{\frac{1}{2}} (\text{time})^{-1}$	$(\text{length})^{\frac{1}{2}} (\text{mass})^{\frac{1}{2}}$

It is possible to overcome this confusion by introducing a fourth fundamental dimension. The quantity conventionally chosen as the fourth dimension is either charge or current. In SI the unit of current (the ampere) is defined and the unit of charge (the coulomb) is a derived quantity. The fourth dimension could however be μ_0 (magnetic constant or magnetic permeability of a vacuum), or ε_0 (electric constant or electric permittivity). When this fourth dimension is introduced, the dimensions of quantities in the ESU, EMU and the non-rationalised MKSA (see below) systems can be expressed uniquely. The same equations can be used in all three systems, but such equations will contain the two-dimensional constants of μ_0 (in H m^{-1}) and ε_0 (in F m^{-1}). Coulomb's law could thus be written in the form

$$F = \frac{Q_1 Q_2}{\varepsilon_0 r^2}$$

where Q_1 and Q_2 are two point charges separated by distance r, and F is the force between charges. The dimensional constants, μ_0 and ε_0 are related through the velocity of light in a vacuum (c) such that

$$\varepsilon_0 \mu_0 = \frac{1}{c^2}$$

in dimensions $\qquad$ F m^{-1}.H m^{-1} = (m s^{-1})$^{-2}$

i.e. $\qquad$ A s V^{-1} m^{-1} . V A^{-1} s m^{-1} = m^{-2} s^2

or $\qquad$ s^2 m^{-2} = m^{-2} s^2.

It has been shown above how it is possible to convert systematically between units and between equations of ESU, EMU and MKSA non-rationalised systems. It is possible to extend this unification to all six common metric systems by the introduction of two more constants of proportionality, c_1 and c_2. In the above systems, electric flux per sphere is equal to $4\pi Q$, where Q is a point charge at the centre of the sphere. (Q = electric charge in coulombs, that is A s). As there are 4π steradians in a sphere, it follows that electric charge per steradian = Q. Now electric flux per sphere can be written as $c_1 Q$, and rationalisation of the systems under consideration requires that values of c_1 be chosen so that 4π will disappear from electrical and magnetic equations where the geometry would suggest its absence, and appear in the cases in which the geometry suggests its presence. Thus $c_1 = 4\pi$ in the ESU, EMU and non-rationalised MKSA systems, but $c_1 = 1$ for the rationalised MKSA system used in SI. Thus the unit of solid angle in the rationalised flux equation is the sphere, while in the non-rationalised flux equation this unit is the steradian. However, it is better to treat the unit of c_1 as distinct from the unit of solid angle, as the solid angle appears in some equations as a variable quantity. In the same way it is also possible to give c_2 a physical interpretation as a measure of circulation linkage between two interlocked loops of wire carrying current. The appearance of c_2 equal to the reciprocal of the velocity of electromagnetic radiation in the symmetrical Gaussian and Heaviside Lorentzian systems is due to setting both ε_0 and μ_0 equal to unity simultaneously.

TABLE 40 NUMERICAL VALUES OF THE CONSTANTS ε_0, μ_0, c_1 and c_2

System	ε_0	μ_0	c_1	c_2
MKSA rationalised	$\approx (1/36\pi \times 10^9)$	$4\pi \times 10^{-7}$	1	1
MKSA non-rationalised	$\approx (1/9 \times 10^9)$	1×10^{-7}	4π	1
EMU	$\approx (1/9 \times 10^{20})$	1	4π	1
ESU	1	$\approx (1/9 \times 10^{20})$	4π	1
Gaussian	1	1	4π	$\approx 1/(3 \times 10^{10})$
Heaviside-Lorentz	1	1	1	$\approx 1/(3 \times 10^{10})$

One can now write the equations in a generalised form. For example,
(i) Coulomb's law in a vacuum:

$$F = \frac{c_1 \, Q_1 Q_2}{4\pi \, \varepsilon_0 \, r^2}$$

or in general:

$$F = \frac{c_1 \, Q_1 Q_2}{4\pi \, \varepsilon_0 \, \varepsilon_r \, r^2}$$

Q_1 and Q_2 are two point charges separated by distance r, and F is the force between the charges. ε_0 is the electric constant (or permittivity of vacuum) and ε_r is the relative permittivity (or dielectric constant) of the dielectric medium.
(ii) Capacitance (C) of an isolated sphere of radius r is given by

$$C = \frac{4\pi \, \varepsilon_r \, \varepsilon_0 \, r}{c_1}$$

(iii) Magnetomotive force (F_m) around n turns of wire each carrying current I

$$F_m = c_1 \, c_2 \, n \, I.$$

It is internationally recommended that the electrical and magnetic equations founded on the four basic quantities of length, mass, time and electric current, and written in rationalised forms, be used. The quantities are expressed in SI units.

APPENDIX II

Expressions of Concentrations of Solutions

The derived physical quantity of concentration in SI is given in the units of moles per cubic metre ($mol\ m^{-3}$). There is at present no accepted name or symbol for a unit equivalent to one $mol\ m^{-3}$.

A large number of methods exist for expressing concentrations of solute in a solvent. We will demonstrate here the relationships between the most common methods and the acceptable SI unit.

Expressions of Concentration in Terms of Weight of Solute in a given Volume of Solution.

Molarity of a solution

A molar solution contains one molecular gramme-weight (1 mole) of the solute dissolved in sufficient solvent to make one cubic decimetre (strictly one litre*) of solution. Thus the molarity of a solution is the expression of the number of moles per cubic decimetre of the solution. It follows that a one molar (often written 1M) solution contains $10^3\ mol\ m^{-3}$.

Formality of a solution

In some cases the molecular weight of the solute is not clearly defined, and in such cases some authors have used the formula gramme-weight of the solute as the measure of weight of substance in the solution. A formal solution thus contains one formula gramme-weight of solute in enough solvent to make one cubic decimetre (strictly one litre) of solution. Where the formula gramme-weight and molecular gramme-weight of a substance are the same, the molarity and formality of a solution are identical.

Normality of a solution

A normal (1N) solution contains one equivalent gramme-weight of the solute in a litre of solution. An equivalent gramme-weight of an acid or base is that weight (in grammes) of the solute that will produce or react with one mole of hydrogen ions; for example, the equivalent weight (E.Wt.) of hydrochloric acid (which produces one mole of H^+ ions per mole of HCl) is equal to the molecular weight. On the other hand, the E.Wt. of $Ca(OH)_2$ (which reacts with 2 moles of H^+ ions per mole of $Ca(OH)_2$) is equal to the molecular weight divided by two.

In reactions in which acid-base terminology cannot be used, such as the precipitation of $BaSO_4$

$$Ba(NO_3)_2 + Na_2SO_4 \rightarrow BaSO_4 + 2NaNO_3$$

the equivalent weights of substances present are given by the molecular weight divided by the number of the total cationic (or anionic) charges on the compound under consideration. Thus

$$\text{E.Wt. of } Ba(NO_3)_2 = \frac{\text{molecular weight}}{2}.$$

* For chemical purposes the litre is very nearly equivalent to one cubic decimetre ($10^3\ cm^3$) in volume. The litre is not an acceptable SI unit.

Effectively the equivalent gramme-weight of such compounds is the weight (in g) of the solute that could produce the corresponding acid anion by reaction with one mole of H^+ ions, viz.

$$Ba(NO_3)_2 + 2H^+ \rightleftharpoons 2HNO_3 + Ba^{2+}$$

or $\frac{1}{2}$ mole of $Ba(NO_3)_2$ reacts with one mole of H^+ ions.

In reduction-oxidation reactions the E.Wt. of a solute is defined in terms of the number of electrons transferred in the redox reaction. For example in acid solution reductions of $KMnO_4$, the oxidation state of Mn(VII) is reduced to Mn(II) by the transfer of five electrons from the species being oxidised.

In such cases the E.Wt. of $KMnO_4 = \dfrac{\text{molecular weight}}{5}$.

In alkaline solution reductions of $KMnO_4$

$$Mn(VII) + 3e^- \rightarrow Mn(IV)$$

i.e.   E.Wt. of $KMnO_4 = \dfrac{\text{molecular weight}}{3}$.

The concept of equivalent weight, and thus normality, is discouraged in most branches of chemistry, as definitions of these concepts can only be offered in terms of a given (or given type of) reaction. The equivalent weight of a compound may change for different reactions. However, analytical chemists still use the concept of normality very widely. As equivalent weights are quantitatively equivalent to each other in a given type of reaction, normalities of solutions expressed on the basis of the reaction are also equivalent. Thus a given volume of solution of given normality will produce or react with the same volume of another solution of the same normality for the specified type of reaction, For instance,

$10dm^3$ of IN HCl will react with and neutralise exactly $10dm^3$ of IN $Ba(OH)_2$.

Expressions of Concentration in Terms of Weight of Solute in a given Weight of Solvent

Molality of a solution

A molal solution contains one molecular gramme-weight (1 mole) of the solute dissolved in one kilogramme of solvent. This is a derived SI unit having dimensions of mole per kilogramme.

Percentage composition of a solution

The weight for weight ($^w/_w$) percentage of a solution is the weight in any given weight unit of the solute dissolved in one hundred similar weight units of solvent. Thus 5g of solute in 100g of solvent represents a 5% solution.

APPENDIX III

Engineering Units

Screw Threads

Since 1841 attempts have been made at various times to get standardisation in screw threads. At the present time over two hundred thread forms exist and about ten of these are used frequently.

Screw threads are classified by their diameter, the number of threads per inch or per centimetre, and their form. The pitch, which is the reciprocal of the number of threads per unit length, is also sometimes given. The form is the shape of one complete profile of the thread between corresponding points at the bottom of adjacent grooves as shown in axial plane section.

In the United Kingdom two basic screw threads are used; namely the British Standard (BS) Whitworth thread which is a coarse thread and the British Standard fine (BSF) thread. In addition the British Association, British Standard Cycle and Unified threads are used extensively. In U.S.A. the oldest system is the American Standard or Franklin screw thread introduced in 1864. Other American screw thread systems include the Society of Automotive Engineers (SAE), and the American Standard Coarse and American Standard Fine. In the U.S.A. and in metric countries the same thread shape, the ISO thread (in both coarse and fine modifications), has been used in recent years. Inch sizes are used in U.S.A. and in the United Kingdom and metric sizes in the rest of the world. Both U.S.A. and British industry are now changing from the inch-based ISO thread to the millimetre-based ISO thread. Some of the large British manufacturers of screws have already announced a higher price for inch-based screws than for metric screws. This price differential will be increased gradually until the inch-based screws cost over 50% more than the equivalent metric screws.

Technical Drawings

The standard scales for technical drawings in the imperial system are:

1 : 1	full size
1 : 4	3 in to the foot
1 : 12	1 in to the foot
1 : 24	$\frac{1}{2}$ in to the foot
1 : 48	$\frac{1}{4}$ in to the foot
1 : 96	$\frac{1}{8}$ in to the foot

The standard metric drawings are:

1 : 1
1 : 5
1 : 10
1 : 100

Maps and Plans

In English-speaking countries large-scale maps are made to the scale 1 : 126 720 ($\frac{1}{2}$ in to the mile) or 1 : 63 360 (1 in to the mile). These scales are replaced by 1 : 100 000 (1 cm to 1 km) and 1 : 50 000 (2 cm to 1 km) respectively.

Town surveys have used the scale 1 : 10 560 (6 in to the mile). The new standard scale is 1 : 10 000 (10 cm to 1 km) so that 1 cm² represents an area of 1 hectare (2·47 acres), and 1 mm² represents 0·01 hectare. The hectare is the standard of land measure in metric countries.

Ordnance survey maps have used the scales 1 : 1 250 and 1 : 2 500. The new scales are 1 : 1 000 (100 cm to 1 km) and 1 : 2 000. House and site plans use the scales 1 : 192 ($\frac{1}{16}$ in to the foot); 1 : 96 ($\frac{1}{8}$ in to the foot) and 1 : 48 ($\frac{1}{4}$ in to the foot). The metric scales are 1 : 200 (5mm to 1m); 1 : 100 (1 cm to 1 m); and 1 : 50 (2 cm to 1 m).

Modules

When industrialised building was introduced the United Kingdom adopted the 4 in module. This is to be changed to the basic metric module of 100 mm for non-structural components such as door frames, cupboards and windows, while the 25 mm module will be used for the thickness of walls, columns and so on. Floor to floor heights will be standardised to 2 500 mm.

APPENDIX IV

Sub- and superscripts

Table 41 lists recommended sub- and superscripts for use in SI.

TABLE 41 SI SUB- AND SUPERSCRIPTS

Subscripts	Refers to
p, k	potential and kinetic energy respectively.
0	a measurement in a vacuum.
m	a molar expression.
r	a relative expression; or a reaction rate; or a rotational expression.
A,B,C ...	substances A, B, C ...
i	a particular number; or a typical ionic species i.
a	an atomic expression.
+ −	a positive or negative ion.
c	a critical state.
d,e,f,s,t	dissolution (or decomposition), evaporation, formation (or fusion), sublimation, transition respectively.
m	a molal expression.
γ	an activity coefficient expression.
x,y,z	mole fractions of a substance.
a	relative activity expression.
V,p,T,S	constant volume, pressure, temperature, entropy respectively.

Superscripts	Meaning
‡	referring to an activated complex.
.	denoting a property of a pure substance.
θ	denoting a standard value of a property.
I,II	denoting different systems, or different states of a system.
id	ideal.
G,L,S	referring to gaseous, liquid or solid states respectively.
E	excess.
∞	indicating a limiting value at infinite dilution.

APPENDIX V

Recommended Values of Physical Constants

Table 42 lists symbols and values for the more important physical constants.

TABLE 42 SYMBOLS AND VALUES OF PHYSICAL CONSTANTS

Symbol	Physical Constant	Value
c_0	speed of light in a vacuum	$(2 \cdot 997\,925 \pm 0 \cdot 000\,003) \times 10^8$ m s^{-1}
μ_0	permeability of a vacuum	$4\pi \times 10^{-7}$ kg m s^{-2} A^{-2}
m_u	unified atomic mass constant	$(1 \cdot 660\,43 \pm 0 \cdot 000\,08) \times 10^{-27}$ kg
m_p	mass of proton	$(1 \cdot 672\,52 \pm 0 \cdot 000\,08) \times 10^{-27}$ kg
m_e	mass of electron	$(9 \cdot 109\,1 \pm 0 \cdot 000\,4) \times 10^{-31}$ kg
e	charge of electron	$(1 \cdot 602\,10 \pm 0 \cdot 000\,07) \times 10^{-19}$ C
N_A	Avogadro constant	$(6 \cdot 022\,52 \pm 0 \cdot 000\,28) \times 10^{23}$ mol^{-1}
F	Faraday constant	$(9 \cdot 648\,70 \pm 0 \cdot 000\,16) \times 10^4$ C mol^{-1}
h	Planck constant	$(6 \cdot 625\,6 \pm 0 \cdot 000\,5) \times 10^{-34}$ J s
k	Boltzmann constant	$(1 \cdot 380\,54 \pm 0 \cdot 000\,18) \times 10^{-23}$ J K^{-1}
R_∞	Rydberg constant	$(1 \cdot 097\,373\,1 \pm 0 \cdot 000\,000\,3) \times 10^7$ m^{-1}
R	gas constant	$(8 \cdot 314\,3 \pm 0 \cdot 001\,2)$ J K^{-1} mol^{-1}
μ_B	Bohr Magneton	$(9 \cdot 273\,2 \pm 0 \cdot 000\,6) \times 10^{-24}$ A m^2
T_{ice}	"ice-point" temperature	$(2 \cdot 731\,500 \pm 0 \cdot 000\,001) \times 10^2$ K
Π	circumference/diameter of a circle	$3 \cdot 141\,592\,65$